Meson and Baryon Spectroscopy

By D. B. LICHTENBERG

Professor of Physics, Indiana University

Augmented, updated edition
of the article in Volume 36 of
SPRINGER TRACTS IN MODERN PHYSICS
Ergebnisse der exakten Naturwissenschaften

SPRINGER-VERLAG · NEW YORK INC.

1965

ISBN-13: 978-0-387-90000-1   e-ISBN-13: 978-1-4615-9636-3
DOI: 10.1007/978-1-4615-9636-3

Reprinted from Ergebnisse der exakten Naturwissenschaften, Band 36.
© by Springer-Verlag, Berlin · Goettingen · Heidelberg 1964.
Library of Congress Catalog Card Number: 25-9130

Contents

Preface

This work originated in a series of lectures on meson and baryon excited states which I gave at the Stanford Linear Accelerator Center in the fall of 1962. The notes of these lectures were issued as a Stanford University report (SLAC–13) in March, 1963. In the fall of 1963, I gave a revised set of lectures on meson and baryon spectroscopy at Indiana University. In both cases, the talks were given primarily for experimental physicists.

In preparing the notes of these talks for publication, I have added some introductory material on pions, nucleons, kaons, and hyperons. My main emphasis is on the experimental facts concerning the spectroscopy of the mesons and baryons and on the use of conservation laws in their interpretation. I have either mentioned briefly or omitted entirely explanations which depend on dynamical models. Although I have given a number of facts about the electromagnetic and weak decays of some mesons and baryons, I have not discussed the theory of these decays.

This is not a comprehensive review of the subject of the strongly interacting particles, and the list of references is not complete. Also, I have not always devoted time to a subject in proportion to its importance, but have spent more time on topics which have particularly interested me. Nevertheless, I hope that I have provided a useful summary of the most important facts about the spectroscopy of the mesons and baryons.

1. Introduction

In the years since the discovery of the pion in 1947, a large quantity of experimental information has been amassed about the strongly interacting particles — the mesons and baryons. With the advent of high energy accelerators, mesons and baryons have been discovered at an increasingly rapid rate. A new spectroscopy has grown up, with the object of determining the masses and other quantum numbers of these particles.

A meson is a strongly interacting particle of integral spin and nonzero mass which decays, either directly or via the intermediary of other mesons, into leptons, photons, or a combination of the two. The known mesons, plus some reported in the literature whose existence is in doubt, are listed in Tables 1.1 and 1.2. A question mark after the symbol for a meson indicates that the evidence in favor of its existence is of a preliminary nature. Our notation for the mesons is the one used most commonly in the literature.

A baryon is a strongly interacting particle of half-odd-integral spin which (except for the proton) decays, either directly or via a chain of decays, into a proton plus lighter particles. The known baryons plus some suspected ones are listed in Tables 1.3, 1.4, and 1.5, the doubtful baryons being indicated by a question mark. Our notation for the baryons

Table 1.1[1]. Mesons with hypercharge $Y = 0$. The symbols used are I (isospin), G (G parity), C (C parity[2]), J (spin), and P (parity). In Tables 1.1 through 1.6, a question mark after the symbol for a particle means either that the evidence in favor of its existence is doubtful, or that a real effect has been observed but the interpretation is doubtful. In Tables 1.1 through 1.5, no distinction has been made among the different kinds of neutrinos

Meson	I^G C J^P	Mass MeV	Width MeV (or mean life, sec)	Principal decay modes	%
π	1^- $+$ 0^-	$\pi^\pm$ 139.6	(2.55×10^{-8})	$\mu^\pm \nu$	≈ 100
		π^0 135.0	(10^{-16})	2γ	98.6
				$\gamma e^+ e^-$	1.4
η	0^+ $+$ 0^-	548	< 7	$\pi^+ \pi^- \pi^0$	30
				2γ	35
				$3\pi^0$	30
				$\pi^+ \pi^- \gamma$	5
ρ	1^+ $-$ 1^-	760	120	2π	≈ 100
ω	0^- $-$ 1^-	784	9	$\pi^+ \pi^- \pi^0$	86
				$\pi^+ \pi^-$	2 ?
				$\pi^0 \gamma$	12
φ	0^- $-$ 1^-	1019	3	$K^+ K^-$	
				$K_1 K_2$	
				3π	< 20
f	0^+ ? $+$? 2^+ ?	1250	120	2π	
B	1^+ $-$?	1220	100	$\pi \omega$	
ABC ?	0^+ ? $+$? 0^+ ?	320	25 ?	2π ?	
$K_1 K_1$?	0^+ ? $+$ 0^+ ?	1020	?	$K^+ K^-$	
				$K_1 K_1$	
ζ ?	1 ? ?	570	?	2π	
$\rho \pi$	?$^-$? 1^+ ?	1200	350	$\rho \pi$	
$K \bar{K} \pi$	0 ? ? ?	1410		$K \bar{K} \pi$	

[1] Most of the numbers given in Tables 1.1 through 1.6 were taken from compilations of ADAIR and FOWLER (1963), BARKAS and ROSENFELD (1963), COHEN and DUMOND (1963), DALITZ (1963), PUPPI (1963), ROOS (1963), and WILSON (1963). The references to the original papers may be found by consulting these works. We have referred to many of the original papers in the sections in which the different particles are discussed. References to data not taken from the compilations may also be found in these sections.

[2] The C parity of a meson of $I > 0$ applies only to the neutral member of the multiplet.

is a modification of that suggested by CHEW, GELL-MANN, and ROSENFELD. A capital Greek or Roman symbol stands for a baryon of particular hypercharge Y and isospin I as follows

$$N \quad \text{for} \quad Y = \quad 1, I = {}^1/_2$$

$$\Delta \quad \text{for} \quad Y = \quad 1, I = {}^3/_2$$

$$\Lambda \quad \text{for} \quad Y = \quad 0, I = 0$$

$$\Sigma \quad \text{for} \quad Y = \quad 0, I = 1$$

$$\Xi \quad \text{for} \quad Y = -1, I = {}^1/_2$$

$$\Omega \quad \text{for} \quad Y = -2, I = 0$$

Table 1.2. Mesons with hypercharge $Y = 1$. The antiparticles are not listed

Meson	I	J^P	Mass MeV	Width MeV (or mean life, sec)	Principal decay modes	%
K	$1/2$	0^-	K± 494	(1.22×10^{-8})	$\mu^\pm \nu$	64
					$\pi^\pm \pi^0$	22
					$\mu^\pm \pi^0 \nu$	3
					$e^\pm \pi^0 \nu$	5
					$\pi^\pm \pi^+ \pi^-$	5
					$\pi^\pm \pi^0 \pi^0$	2
			K_1 498	(1.0×10^{-10})	$\pi^+ \pi^-$	70
					$2\pi^0$	30
			K_2 498	(6×10^{-8})	$\mu^\pm \pi^\mp \nu$	30
					$e^\pm \pi^\mp \nu$	33
					$\pi^+ \pi^- \pi^0$	13
					$3\pi^0$?	21
K*	$1/2$	1^-	888	50	$K\pi$	
ϰ ?	$1/2$	?	726	20 ?	$K\pi$	
$K\pi\pi$?	$1/2$?	?	1170	< 50	$K\pi\pi$	
$K\pi\pi$?	$1/2$	1^+ ?	1230	80	$K\pi\pi$	
					$K\rho$	

Table 1.3. Baryons with hypercharge $Y = 1$ (strangeness 0). The antiparticles are not listed

Baryon	I	J^P	Mass MeV	Width MeV (or mean life, sec)	Principal decay modes	%
N	$1/2$	$1/2^+$	p 938.2	stable		
			n 939.5	(1.01×10^3)	$pe^- \nu$	100
Δ	$3/2$	$3/2^+$	1238	90	$N\pi$	≈ 100
N (1512)	$1/2$	$3/2^-$?	1512	100	$N\pi$	
					$N\pi\pi$	
N (1688)	$1/2$	$5/2^+$?	1688	100	$N\pi$	
					$N\pi\pi$	
					ΛK ?	
					$N\eta$?	
Δ (1920)	$3/2$	$7/2^+$?	1920	200	$N\pi$	
					$N\pi\pi$	
					ΛK ?	
N (2190)	$1/2$	?	2190	200		
Δ (2360)	$3/2$	?	2360	200		
N (1700) ?[1]	$1/2$	$1/2^+$?	1700	?	πN	
					ΛK	
Δ (1650) ?	$3/2$		1650	?	πN	

[1] The N (1700) may be the same particle as the N (1688).

These symbols without any additional notation stand for the lowest mass states with these quantum numbers. An excited state is indicated by the mass in MeV in parentheses, or, where no ambiguity will arise in context, by an asterisk after the symbol. An excited state of hypercharge zero and unspecified isospin is given the symbol Y^*. Sometimes the older notation is also given.

It is quite likely that many more mesons and baryons exist which have not yet been discovered. Also, it is possible that a few of the particles listed as known will turn out not to exist after all, although the

Table 1.4. Baryons with $Y = 0$. The antiparticles are not listed

Baryon	I	J^P	Mass MeV	Width MeV (or mean life, sec)	Principal decay modes	%
Λ	0	$1/2^+$	1115.4	(2.5×10^{-10})	$p\pi^-$ $n\pi^0$	64 36
Σ	1	$1/2^+$	Σ^+ 1189.4	(0.8×10^{-10})	$p\pi^0$ $n\pi^+$	50 50
			Σ^0 1193.2 Σ^- 1197.6	$(<10^{-11})$ (1.6×10^{-10})	$\Lambda\gamma$ $n\pi^-$	≈ 100 ≈ 100
Σ (1385)	1	$3/2^+$	1385	50	$\Lambda\pi$ $\Sigma\pi$	≈ 98 <4
Λ (1405)	0	$1/2^-$?	1405	40	$\Sigma\pi$ $\Lambda\pi\pi$	
Λ (1520)	0	$3/2^-$	1520	16	$\Sigma\pi$ $\overline{K}N$ $\Lambda\pi\pi$	55 30 15
Σ (1660)	1	$3/2^+$?	1660	40	$\overline{K}N$ $\Sigma\pi$ $\Lambda\pi$	
Λ (1815)	0	$5/2$?	1815	120	$\overline{K}N$ $Y\pi$	

Table 1.5. Baryons with $Y = -1$ and -2. The antiparticles are not listed

Baryon	I	J^P	Mass MeV	Width MeV (or mean life, sec)	Principal decay modes	%
Ξ	$1/2$	$1/2^+$?	Ξ^0 1316 Ξ^- 1321	$(\approx 10^{-10})$ (1.3×10^{-10})	$\Lambda\pi^0$ $\Lambda\pi^-$	≈ 100 ≈ 100
Ξ (1530)	$1/2$	$3/2^+$?	1529	7	$\Xi\pi$	100
Ω	0	?	1685	$(\approx 10^{-10})$	$\Xi\pi$ $\Lambda\overline{K}$	

evidence in favor of them at present seems fairly convincing. Some additional possible mesons and baryons, supported by inconclusive evidence and not listed in Tables 1.1 through 1.5 or discussed in the text, are given in a review by Roos (1963).

For completeness, we list in Table 1.6 the known leptons (particles of half-odd-integral spin which do not participate in strong interactions), the photon (the quantum of the electromagnetic field), the weak boson

Table 1.6. Other particles

Particle	Symbol	Mass MeV	Spin	Mean life sec	Observed decay modes
Photon	γ	0	1	stable	—
Electron	$e\pm$	0.511	$1/2$	stable	—
Muon	$\mu\pm$	106	$1/2$	2.20×10^{-6}	$e\pm\nu\bar{\nu}$
e-neutrino	$\nu_e, \bar{\nu}_e$	0	$1/2$	stable	—
μ-neutrino	$\nu_\mu, \bar{\nu}_\mu$	0	$1/2$	stable	—
Graviton ?		0	2	stable	—
Weak boson ?	$W\pm, W^0$? $\overline{W}^0$?	1300 ?	1 ?	?	

(quantum of the weak interaction), and the graviton (quantum of the gravitational field). At present, the experimental evidence in favor of the weak boson is of a preliminary nature, and the graviton has not been experimentally detected, presumably because of the weakness of its interaction.

Masses, lifetimes, and other constants, in the tables and in the text, are usually taken from one of the following sources: ADAIR and FOWLER (1963), BARKAS and ROSENFELD (1963), COHEN and DUMOND (1963), DALITZ (1963), PUPPI (1963), ROOS (1963) and WILSON (1963). See these reviews for the original references*. When our information has come from another source, we have endeavored to give the reference; our apologies to the original authors if we have not always succeeded.

We have listed in the references a number of books on elementary particles and related topics, even though we have not had occasion to refer to them. With the aid of these books, the reader can go more deeply into topics of interest which have been discussed only briefly in these notes.

2. Conservation laws

A. Exact and approximate interaction symmetries

Our theoretical understanding of the mesons and baryons has not kept pace with the accumulation of experimental facts. There is not yet a quantitatively successful dynamical theory of the interactions of these particles. Nevertheless, much of the experimental information has been correlated by the use of a number of symmetries or invariance principles which appear to hold in elementary particle interactions.

Usually, an interaction symmetry or invariance principle has as a consequence a conservation law. An exception is time reversal invariance. Since the time reversal operation is antiunitary, invariance of an interaction under this operation leads to no conservation law (WIGNER 1959). However, time reversal invariance does have experimentally verifiable consequences.

There are both exact and inexact conservation laws. An approximate conservation law is recognizable if the interactions in which the law is violated are weaker than those in which it holds. Inexact conservation laws have been observed because of a hierarchy in the strengths of the different kinds of interactions: strong, electromagnetic, weak, and gravitational. Since the strong and weak interactions are not well understood, the terms "strong" and "weak" in this context should be interpreted as generic words which may each correspond to more than one kind of interaction. There may in addition be still other kinds of interactions which have not been experimentally detected. A preliminary

---

* When a reference appears in parentheses, only the first-named author is given; e.g. ADAIR and FOWLER (1963) but (ADAIR 1963).

experiment by LEIPUNER et al. (1963) involving K mesons has in fact been interpreted by the authors as possible evidence for a "fifth force".

The gravitational force appears too weak to have any measurable effect on elementary particle interactions at presently available particle energies. Although gravitational forces may somehow be intimately connected with particles and their interactions, no one has yet demonstrated a connection that has observable consequences. We shall not consider the gravitational interaction further.

Some important conservation laws and invariance principles which, as far as we know, are exact are conservation of momentum and energy, conservation of angular momentum, conservation of charge and charge number, conservation of baryon number, conservation of two lepton numbers (muon number and electron number), and time reversal invariance. Some of these laws will be discussed in more detail later.

The concept of an approximate law is in some respects peculiar, because no matter how weakly a law is violated in most reactions, there may occur reactions or decays in which the extent of the violation is so large that the law cannot be recognized. Some inexact conservation laws and invariance principles are conservation of hypercharge (or alternatively strangeness), conservation of isospin* and its z-component, conservation of parity, and charge conjugation invariance (or conservation of C parity). All these laws are violated in weak interactions, and, in addition, isospin is not conserved in electromagnetic interactions. However, the conservation law corresponding to taking the product of parity and charge conjugation is, as far as is known, exact.

There may be other inexact interaction symmetries which are broken even in strong interactions and are not readily discernible. Examples of such symmetries which have been proposed in the literature are global symmetry (SCHWINGER 1956, 1957, GELL-MANN 1957), cosmic symmetry (SAKURAI 1959), and unitary symmetry (SAKATA 1956, GELL-MANN 1961, 1962, NE'EMAN 1961). Still other possible conservation laws and symmetry principles have been discussed. Some of them are symmetries of weak but not strong interactions. Thus, it is not clear whether the often-made statement, "The stronger the interaction, the higher the number of symmetries," is true. However, a symmetry of weak interactions which is not shared by the strong interactions does not, in general, show up as an approximate conservation law.

It may be that there are still other conservation laws whose existence we have not yet suspected, or that some we have mentioned may prove not to be useful when we look at things in a different way. We shall say little about such possibilities.

B. Quantum numbers of elementary particles

For each conservation law, corresponding to the invariance of an interaction under a particular transformation, there exists the possi-

---

* Isospin is often called isotopic spin, isobaric spin, or I spin.

bility of assigning a quantum number to an elementary particle. This can be done if the state function of the particle is an eigenstate of the operator representing the transformation.

An elementary particle is characterized by a mass M and lifetime τ, a spin J, a charge number Q, a baryon number B, and two lepton numbers L_μ and L_e. A baryon has $B = 1$, $L_\mu = L_e = 0$; a meson has these three quantum numbers all zero. The mesons and baryons have additional quantum numbers, parity P, isospin I and z-component I_z, and hypercharge Y (or alternatively strangeness S). Mesons with $Y = 0$ have still another quantum number, G parity, and mesons with $Y = Q = 0$ have the quantum number, C parity. It has been suggested that mesons and photons have yet one more quantum number, A parity (BRONZAN 1963).

The quantum numbers of a particle can be divided into two classes — those having to do with its space-time or "external" properties, such as spin and parity, and those having to do with properties in some "internal" space, such as baryon number and isospin. In general, it is easier to measure the internal quantum numbers of a particle than its external ones. This is because the external quantum numbers of a particle often cannot be obtained without a knowledge of its orbital motion. By definition, there is no such coupling between the internal quantum numbers of a particle and its external motion. Thus, for example, the isospin of a particle is usually measured more easily than its spin and parity.

An additive quantum number is one in which the total quantum number specifying the state of a group of particles is determined by adding up the quantum numbers of the individual particles (including, where relevant, the quantum numbers associated with the external motion). Some additive quantum numbers are Q, B, Y, L_μ, L_e, and J_z. Conservation of the internal quantum numbers (all these except J_z) are associated with so-called gauge transformations of the first kind (ROMAN 1961). A conservation law related to the conservation of Q is the conservation of the magnitude of the electric charge, which is related to invariance under a gauge transformation of the second kind. The conservation of the magnitude of the electric charge is associated with a conserved vector current and with a massless vector field to carry it (the electromagnetic field).

It is not known whether there exist analogous laws for "baryonic" and "leptonic" charges. YANG and MILLS (1954) have suggested that there may exist such a vector field associated with isospin conservation. More recently, SAKURAI (1960) extended this idea and suggested that vector mesons should be associated with baryon, hypercharge, and isospin conservation. Subsequently, vector mesons which might possibly have the desired properties have been discovered, and will be discussed later. However, the analogy between the photon and the vector mesons may not be a correct one because of the following circumstances:

1. The vector mesons have mass, whereas the photon does not.

2. The conservation laws of isospin and hypercharge are not exact (although the conservation of baryon number appears to be).

3. At present we do not have an operational way of obtaining precise measurements of the magnitudes of the baryonic charge, isospin charge, and hypercharge.

The analogy with electromagnetism was also extended to the weak interactions by GERSTEIN and ZELDOVICH (1955) and by FEYNMAN and GELL-MANN (1958) who hypothesized that the vector part of the weak interaction is conserved. Some experimental evidence in favor of this hypothesis was subsequently found by LEE, MO, and WU (1963) in connection with nuclear beta decay, and by several groups (DUNAITSEV 1962, BACASTOW 1962, DEPOMMIER 1963) in connection with pion beta decay.

A multiplicative quantum number is one in which the total quantum number specifying a system of particles is determined by multiplying the quantum numbers of the particles together (including, where relevant, the quantum numbers specifying the external motion of the particles). Parity, C parity, and G parity are examples.

Conservation of parity in an interaction arises from the invariance of the interaction under the replacement $r \to -r$. As far as is known, the strong interactions of mesons are consistent with the assumption that every meson has a definite intrinsic parity, either positive or negative. On the other hand, the wave function of a fermion satisfying the free Dirac equation is not an eigenfunction of the parity (SCHWEBER 1961); i.e., $\Psi(-r) \neq \pm \Psi(r)$. Nevertheless, if a baryon takes part in an interaction which conserves parity, it is permissible to regard the baryon as having a definite intrinsic parity. This follows because of the conservation of baryon number.

It is possible to regard a multiplicative quantum number as an additive quantum number modulo 2. In this way of looking at things, for example, all positive parity particles are assigned "additive parity" zero, and negative parity particles are assigned "additive parity" unity. This procedure leads to selection rules which are completely equivalent to those obtained when parity is regarded as a multiplicative quantum number. The only reason for bringing this up is because there exists the formal possibility for additive quantum numbers modulo 3, 4, 5, ... Thus, for example, there is no evidence on the question of whether hypercharge is a true additive quantum number or whether it is additive modulo some number, say 5. However, since there is at present no theoretical or experimental reason for doubting the additivity of hypercharge, we shall assume that it is a true additive quantum number.

C. Isospin and hypercharge

The mesons and baryons are observed to occur in multiplets. Different members of the same multiplet are identified because they have approximately equal masses and differ in charge number by unit steps. Also, the interactions of different members of a multiplet do not appear to depend strongly on their charges.

For these reasons, it is useful to regard each member of a given multiplet as a different charge state of a single particle, which can be considered to have an extra degree of freedom in an internal space, isospin space. The number of possible orientations of a particle in isospin space is $2I + 1$, if I is the isospin quantum number of the particle. Thus, the isospin quantum number of a particle may be determined simply by finding out in how many charge states it can be produced.

There exists an analogy betweeen isospin and ordinary spin, but it is not exact. We mention the following differences:

1. Whereas any direction may serve as the z axis in ordinary space, a particular direction is singled out in isospin space, since I_z is related to the charge number Q. Since the charge cannot be turned off, isospin must be only approximately conserved. In most cases in which isospin conservation has been tested, it has been found to hold to 3% or better. However, it has not been tested for all the strongly interacting particles.

2. Whereas all rotations in ordinary space are physically meaningful, only certain finite rotations in isospace have physical significance. This is because every observed reaction occurs between states of definite charge which are integral multiples of the unit charge. Thus, only a rotation which takes the initial state to a state of charge number nQ, where n is an integer, is physically meaningful*.

States differing by one unit in I_z differ by one unit in Q. It is found experimentally, however, that I_z does not necessarily equal Q, but can differ from it by a constant which is characteristic of a particular multiplet. It is therefore useful to introduce an additional quantum number, hypercharge, for each particle to measure the difference between Q and I_z for each member of the multiplet. The hypercharge Y of a particle is defined by

$$Q = I_z + \tfrac{1}{2}Y . \tag{2.1}$$

GELL-MANN (1953) and NAKANO and NISHIJIMA (1953) first discovered this rule. GELL-MANN, however, did not use the quantum number Y, but rather the strangeness S, defined by

$$Y = S + B . \tag{2.2}$$

The quantum numbers Y and I_z are conserved in electromagnetic as well as strong interactions, but are not conserved in weak interactions.

D. Charge conjugation and C parity

It appears to be a fact of nature that for every particle with charge number Q, baryon number B, lepton numbers L_e, L_μ, hypercharge Y, and z component of isospin I_z, there is an antiparticle with the same mass

---

* For this reason, it may be better to think of representations of the isospin group as representations of SU_2 rather than of the rotation group.

and lifetime, having quantum numbers $-Q$, $-B$, $-L_e$, $-L_\mu$, $-Y$, and $-I_z$. (If the particle has a magnetic moment or higher moments, these moments are opposite in sign for particle and antiparticle, since the moments are proportional to the charge.) The other quantum numbers of the antiparticle (except for the parity in the case of a fermion) are the same as those of the particle.

If neutrinos have only two components, as is currently believed, then a neutrino is characterized by an additional quantum number, the component of the spin in the direction of motion, or helicity. The helicity of an antineutrino is opposite to that of a neutrino.

By definition, a "particle" has B, L_e, and L_μ positive or zero. If these are all zero, a particle has Y positive or zero. If Y is also zero, there is no convention about which member of a particle-antiparticle pair is the particle. For example, both the π^+ and π^- are regarded as particles (together with the π^0 they form an isospin triplet), even though each is the antiparticle of the other.

Consider an operator C, called the charge conjugation operator, which replaces the state function of a particle with quantum numbers Q, B, L_e, L_μ, Y, I_z by the state function of a particle with quantum numbers $-Q$, $-B$, $-L_e$, $-L_\mu$, $-Y$, $-I_z$. The operator C, acting on the state function of a baryon, meson, or charged lepton, produces the state function of an antibaryon, antimeson, or antilepton respectively. To denote an antiparticle, we write a bar over the symbol for a particle. For most particles X, we have $CX = \overline{X}$.

The strong and electromagnetic interactions of a collection of particles are, as far as is known, identical to the corresponding interactions of the same collection of antiparticles, but the weak interactions are different. This means that C is an approximate symmetry operation.

However, C operating on a neutrino state with lepton number L replaces this state by a state with lepton number $-L$ and the same helicity as the helicity of the neutrino. But if the two-component theory of the neutrino (LANDAU 1957, LEE, 1957, SALAM 1957) is correct, the antineutrino has helicity *opposite* that of the neutrino. Thus, C acting on a neutrino is undefined. Since P reverses the helicity of a particle, the combined operation CP replaces a neutrino by an antineutrino. Thus we write $\overline{\nu} = CP\nu$. It is also sometimes convenient to define a $\overline{K}^0$ meson as CPK^0 rather than CK^0. We shall have occasion to do this in Section 10E on the decays of neutral kaons.

LÜDERS (1959) defines an antiparticle state $\overline{X}$ as $(CPT)X$, where T is the time reversal operator. Such a definition has certain advantages, as can be seen from the paper of LÜDERS. However, there is the disadvantage that T is antiunitary, and thus so is CPT.

The conservation law of baryon number involves antibaryons as well as baryons. It states that the number of baryons minus the number of antibaryons is a conserved quantity. Thus, baryons and antibaryons can be created or destroyed only in baryon-antibaryon pairs. Similar remarks hold for lepton number conservation.

An empirical rule for the mass differences within isospin multiplets is that the more negatively charged the member of the multiplet, the higher is its mass. We mention this rule in connection with particle-antiparticle pairs in order to point out an exception. If a meson has hypercharge zero and isospin greater than zero, the members of the multiplet having charge Q and $-Q$ are antiparticles of each other and have the same mass.

If a particle has quantum numbers

$$Q = B = L_e = L_\mu = Y = 0 , \tag{2.3}$$

then it is indistinguishable from its antiparticle, unless there are as yet undiscovered quantum numbers which are different for particle and anti-particle. If the particle and antiparticle are indistinguishable, then the state function of the particle is an eigenstate of C with possible eigenvalues $C = \pm 1$. Such a particle is said to have a definite C parity. Any collection of particles and antiparticles with total quantum numbers satisfying Eq. (2.3) may also have a definite C parity, but it is not simple in general to determine whether the C parity of such a system is positive or negative.

According to quantum electrodynamics, the photon has $C = -$. This can be seen as follows: The interaction between photons and charged particles is $j_\mu A_\mu$ where j_μ is the current and A_μ is the vector potential. But j_μ changes sign under C because the charge changes sign. Therefore, since the interaction is invariant under C, A_μ must also change sign. But a photon is a quantum of the field A_μ, and therefore must change sign under C if A_μ does. Therefore, the photon is odd under C. A collection of n photons has C parity given by

$$C = (-1)^n . \tag{2.4}$$

E. G conjugation and G parity

For a strongly interacting particle to be an eigenstate of the charge conjugation operator (i.e., to have a definite C parity) it must have quantum numbers $Q = B = Y = 0$. A meson with $Y = 0$, but with $I > 0$ is a member of a multiplet with $2I + 1$ particles, only one of which has $Q = 0$. Therefore only the neutral member of such a multiplet is an eigenstate of C.

However, all members of the multiplet can be shown to be eigenstates of an operator G, called G conjugation (MICHEL 1952, PAIS 1952a, LEE 1956a, GOEBEL 1956), which is defined by

$$G = C R_y , \tag{2.5}$$

where R_y denotes a rotation of $180°$ around the y-axis in isospin space. The eigenvalues of G are $G = \pm 1$, and an eigenstate of G is said to have a definite G parity.

For a particle to be an eigenstate of G, it must have $Y = B = 0$. This follows because C acting on a state with quantum numbers Q, B

and Y replaces this state by the antiparticle state with quantum numbers $-Q$, $-B$ and $-Y$. But a rotation in isospin space can change the charge of a particle but not Y or B. Therefore, unless $Y = B = 0$, an isospin rotation following operation with C cannot lead to the original state.

Two useful relationships follow from the definition of G.

1. If a state consists of several mesons (which need not necessarily be identical) each of which is an eigenstate of G, the G parity of the state is the product of the G parities of the individual mesons.

2. If a neutral meson has a definite isospin and C parity, its G parity is given by

$$G = C(-1)^I. \qquad (2.6)$$

This formula holds for any system which is an eigenstate of C and I.

It may be of interest to ask whether a meson of zero hypercharge is necessarily an eigenstate of G. If it is not, it can be written as a linear combination of two mesons, one with $G = +$ and the other with $G = -$. These two new mesons will have different interactions, and since these interactions are in general strong, the mesons with $G = +$ and $G = -$ may have masses which differ appreciably. Also they will probably be produced in different reactions. If so, there seems to be no point in identifying them as a single meson which is not an eigenstate of G.

F. Theorems from field theory

At present, physicists are not able to use field theory to obtain agreement with experiment when they make dynamical calculations of the interactions of mesons and baryons. Nevertheless, a number of very useful theorems about quanta satisfying the postulates of conventional field theory have been proved. Arguments have been given (STAPP 1963) that the theorems also follow from the S-matrix (scattering matrix) theory, but the derivations from field theory are more rigorous. The postulates of field theory which are necessary to prove the theorems are

1. The interactions of the fields are invariant under proper Lorentz transformations.

2. The fields are local.

3. The fields either commute or anticommute at space-like distances.

4. The vacuum is the lowest energy state.

5. The state vectors form a Hilbert space with positive definite metric.

A further discussion of these postulates is given in the book by SCHWEBER (1961).

We quote without proof the two most important theorems which follow from the postulates.

1. Identical particles of integral spin obey Bose statistics; i.e., any wave function describing these particles is symmetric under the interchange of the coordinates of any two of them. On the other hand,

identical particles of half-odd integral spin obey Fermi statistics; i.e., the wave function of the particles is antisymmetric under the interchange of the coordinates of any two of them (LÜDERS 1958, BURGOYNE 1958).

2. The product of time reversal, charge conjugation, and parity, is an exact symmetry-operation (PAULI 1955, LÜDERS 1957, JOST 1957). This theorem is known as the TCP (or CPT) theorem.

Suppose, in addition, the free fields (without interaction) of spin 0 particles satisfy the Klein-Gordon equation, the free fields of spin $1/2$ particles satisfy the Dirac equation, and the free fields of spin 1 particles satisfy the Proca equation. Then the following two additional theorems can be proved* about the quanta of these fields:

3. If a boson-antiboson pair participate in parity conserving reactions so that their intrinsic parities can be defined, then the parity of the pair in a state of orbital angular momentum L is

$$P = (-1)^L . \tag{2.7}$$

On the other hand, under the same circumstances the parity of a fermion-antifermion pair is given by

$$P = (-1)^{L+1} . \tag{2.8}$$

4. If a particle-antiparticle pair participate in interactions which are invariant under C, and if the pair are in a state of orbital angular momentum L and total spin S, then the pair have a definite C parity given by**

$$C = (-1)^{L+S} . \tag{2.9}$$

This relationship holds for both fermion-antifermion and boson-antiboson pairs.

Theorems 3 and 4 hold not only for particles of spin 0, $1/2$, and 1, but for any particles which are bound states or resonances of such particles.

Theorem 1 refers to the symmetry of a wave function under the interchange of all the coordinates of any two identical particles. This theorem can be extended to include different members of the same isospin multiplet by including the isospin coordinates of the particles. If the theorem is valid, two identical baryons or mesons in a state of angular momentum L, spin S, and isospin I satisfy the equation

$$(-1)^{L+S+I+1/2 Y} = 1 . \tag{2.10}$$

From Theorem 2 (the CPT Theorem), it can be deduced (LEE 1956, 1957, LÜDERS 1957a) that the mass and lifetime of a particle are the same as that of its antiparticle. However CP invariance alone is sufficient to prove this (SACHS 1963). From Theorem 4 and the relationship $G = (-1)^I C$, we obtain for a baryon-antibaryon or meson-antimeson pair

$$G = (-1)^{L+S+I} . \tag{2.11}$$

---

* It may be possible to prove these two theorems without these added restrictions on the free fields.

** We use the symbol J for the spin of a particle or the total angular momentum of a system, and the symbol S for the spin of a two-particle system. The spin S is not necessarily a good quantum number.

Eq. (2.11) holds even if the total charge of the pair is not zero, provided the pair is in a definite state of G, L, S, and I.

It is not obvious that real particles obey the postulates of field theory on which these theorems are based. Therefore a number of people have questioned these results and suggested experimental tests.

GENTILE (1940) and GREEN (1953) have considered fields which neither commute nor anticommute at space-like distances but obey more complicated commutation rules. The particles of such fields are called parabosons or parafermions.

JAUCH (1960) and others have argued that parastatistics are incompatible with the existence of a complete set of commuting observables. Recently GREENBERG and MESSIAH (1963) have made a detailed study of the question of the experimental evidence concerning the statistics of the known particles. They came to the conclusion that although electrons and nucleons are indeed fermions and photons are bosons, the evidence concerning most other particles is inconclusive. These authors proposed a number of tests of the statistics of mesons.

FOLDY (1956) has considered Lorentz invariant theories in which the parity and C parity of a particle-antiparticle pair do not necessarily satisfy the usual relationships. Unusual parity and C parity assignments were also considered by SHIROKOV and OKONOV (1960) and SHIROKOV (1961), and these authors proposed experimental tests.

SACHS (1963) has questioned the validity of the CPT theorem and had suggested experimental tests. MEYER et al. (1963) have made a precise measurement of the lifetime of the μ^+ and μ^-. They found

$$\tau_{\mu^+} = (2.197 \pm 0.002) \times 10^{-6} \sec ,$$
$$\tau_{\mu^-} = (2.198 \pm 0.002) \times 10^{-6} \sec ,$$
$$\tau_{\mu^-}/\tau_{\mu^+} = 1.000 \pm 0.001 .$$

This is good evidence that either CP or CPT is conserved.

It is important to emphasize that in every case in which the theorems of field theory have been tested they have been found to agree with experiment. Furthermore, there seems no experiment which would be better understood if any of the results were false. For these reasons, we shall make use of these results even in cases where they have not been tested.

3. Methods of obtaining information about the spins of particles

A. Pure and mixed states, polarization, alignment, helicity

A particle may be produced in a reaction in either a pure or mixed spin state. The particle is said to be in a pure state if its spin wave function ψ_J is given by

$$\psi_J = \sum_{m=-J}^{J} a_m \chi_J^m , \tag{3.1}$$

where the χ_J^m are spin eigenfunctions with respect to some fixed direction in space and the a_m are, in general, functions of the production angle. Since the a_m are complex, they are specified by $2(2J+1)$ real numbers. However, since an overall phase is unobservable, a pure state is specified by $4J+1$ real numbers (or, if the state is normalized, by $4J$ real numbers).

If the state of a particle is an incoherent sum of pure states, then the particle is said to be in a mixed state. All the information about the mixed state can be obtained from a knowledge of a density matrix, a Hermitian matrix with $2J+1$ rows and columns. A discussion of the density matrix is given in the review article by FANO (1957). Since the density matrix is Hermitian, a knowledge of $(2J+1)^2$ real numbers is required to specify a mixed state completely (or, if the state is normalized, then $(2J+1)^2 - 1$ numbers).

Often, however, one must be satisfied with less than a complete knowledge of the state. One particularly useful piece of information is a knowledge of the polarization P of the state, defined (in the rest system of the particle) by

$$P = (\psi_J, \, J\psi_J)/[J\,(\psi_J, \, \psi_J)], \quad J \neq 0 \tag{3.2}$$

for a pure state and by

$$P = \sum_i (\psi_{J_i}, \, J\,\psi_{J_i})/[J\,(\psi_{J_i}, \, \psi_{J_i})] \tag{3.3}$$

for a mixed state, where the notation (ψ_J, ψ_J) indicates a sum over spins, but not an integration over angles. Since the spin is an axial vector, so is the polarization.

Unpolarized particles may have their spins aligned. If the population of states with z-component of the spin m equals the population with z component $-m$, but all m-values are not equally populated, then the spins are said to be aligned. The importance of polarization and alignment for determining the spins of particles is that unstable particles which are unpolarized and unaligned decay isotropically.

The term polarization is often used with another meaning, expecially when applied to integral spin particles. A particle of spin J may be represented by a tensor of rank J which is called a polarization tensor. For example, a photon ($J = 1$) is represented by a polarization vector (a true vector) which must be perpendicular to the direction of motion of the photon. When we say a photon is linearly or circularly polarized, we are referring to this transverse polarization vector. A right-handed circularly polarized photon has its axial vector polarization (i.e., expectation value of its spin), *along* its direction of motion.

The concept of the component of spin along the direction of motion of a particle is a useful one; such a component is called the helicity of the particle. A right-circularly polarized photon has helicity $+1$, and a left-handed photon has helicity -1. The helicity of a particle, unlike the component of its spin along some axis fixed in space, is invariant under rotations. JACOB and WICK (1959) have discussed scattering processes in terms of helicity eigenstates.

B. Production of particles

Consider the reaction

$$a + b \rightarrow c + d \qquad (3.4)$$

where the incident and target particles are unpolarized and unaligned. We shall quote some theorems which limit the magnitude of the production cross section and the complexity of the production angular distribution[*] $I(\Theta, \Phi)$. We also discuss limitations on the polarization of the produced particles. We denote by $I(\Theta)$ the angular distribution averaged over Φ.

1. The magnitude of the contribution to the total cross section in any state of angular momentum J is limited by

$$\sigma_J^2 \text{ (total)} \leqq 4\pi \lambda^2 (2J + 1) \, \sigma_J \text{ (elastic)} \qquad (3.5)$$

where $2\pi\lambda$ is the wave length of the incident particle in the center of mass system. This inequality holds if particles a and b have spin 0. If one particle has spin $^1/_2$ and the other has spin 0, then the cross section in any state of angular momentum and parity J^P is limited by

$$\sigma_J^2 \text{ (total)} \leqq 4\pi \lambda^2 (J + ^1/_2) \, \sigma_J \text{ (elastic)} \qquad (3.6)$$

if parity is conserved in the production process. These limits can be derived from the unitarity of the S-matrix describing the collision[**] (see for example BLATT, 1952). These inequalities may be useful to show that the spin of a resonance is greater than a certain value.

2. If c and d are produced with a maximum angular momentum J and if the polarization of c and d are not measured, then the highest power of $\cos\Theta$ that can occur in the production angular distribution $I(\Theta, \Phi)$ is $\cos^{2J}\Theta$. This theorem was demonstrated by YANG (1948) by making use of rotation matrices. We omit the proof. The theorem is useful because in some cases a and b form a resonant state which then decays into c and d. The theorem gives information on the spin of the resonance.

3. If parity is conserved in the production reaction, then if c and d are produced polarized, the direction of polarization must be perpendicular to the production plane. This can be seen by assuming the polarization is in the production plane and looking at the interaction in a mirror. The reflected reaction is inequivalent to the original reaction although the initial states of the original and reflected reactions are the same. This shows that parity is not conserved, in contradiction to assumption. The theorem is useful because it is often advantageous to analyze the decay of an unstable particle with respect to its direction of polarization.

4. If parity is conserved in the production reaction and if c and d are produced in the forward-backward direction then they are produced unpolarized. This follows because the production plane degenerates into

---

[*] We use capital Θ, Φ to denote production angles, Θ being measured with respect to the incident beam and Φ with respect to any fixed axis, and lower case θ, φ to denote decay angles measured with respect to any axes not depending on the decay.

[**] The unitarity of the S-matrix is a consequence of the conservation of probability.

a line and gives no preferred direction for the polarization. Thus, to maximize the polarization of produced particles, one should reject events produced near the forward direction. Particles produced in the forward direction may have their spins aligned, however.

C. Decays of particles

Information can be obtained about the spin of an unstable particle from its decay angular distribution. Suppose a particle of spin J decays into two other particles. We shall quote some theorems about the decay angular distribution $I(\theta, \varphi)$, where θ, φ are measured with respect to any axes not depending on the decay.

1. The highest power of $\cos\theta$ which can occur in $I(\theta)$ is $\cos^{2J}\theta$. The proof is similar to the one limiting the complexity of a production angular distribution (YANG 1948).

2. If the particle is unpolarized and unaligned, the decay is isotropic. This result follows immediately from the fact that there is no preferred axis in the problem.

3. If parity is conserved in the decay, only even powers of $\cos\theta$ are contained in $I(\theta)$ when averaged over φ. The highest power of $\cos\theta$ appearing in $I(\theta)$ is $\cos^{2J}\theta$ if J is integral, or $\cos^{2J-1}\theta$ if J is half-odd integral.

4. A particle of spin unity cannot decay into two photons. The proof, given by YANG (1950), depends on the fact that the polarization vector of a photon must be transverse.

5. If a particle decays into a spin 0 particle and a spin $1/2$ particle with non-conservation of parity and has a decay angular distribution

$$I(\theta) = 1 + \alpha \cos\theta , \qquad (3.7)$$

then the magnitude of α may give information about the spin of the decaying particle. Specifically, LEE and YANG (1958) showed that the following inequality holds between α and the spin J of the particle

$$-\frac{1}{6J} \leqq \frac{\alpha}{3} \leqq \frac{1}{6J} . \qquad (3.8)$$

More generally, if the particle decays with an angular distribution $I(\theta)$, then

$$-\frac{1}{2J+2} \leqq \langle\cos\theta\rangle \leqq \frac{1}{2J+2} , \qquad (3.9)$$

where

$$\langle\cos\theta\rangle = \int_{-1}^{1} I(\theta) \cos\theta \, d(\cos\theta) / \int_{-1}^{1} I(\theta) \, d(\cos\theta) .$$

We omit the proofs of these theorems.

EBERHARD and GOOD (1960) and PESHKIN (1961, 1963) have proposed theorems which are somewhat more complicated than those given by LEE and YANG, but which may yield more information about the spin.

All these theorems apply only to a particle which decays freely. If the lifetime of a particle is so short that it decays in the interaction region in which it is produced, the problem of determining the spin becomes quite complicated. The appearance of an odd power of $\cos\theta$ in $I(\theta)$ means that the particle does not decay freely (if parity is conserved in the decay). Of course, the absence of an odd power of $\cos\theta$ does not necessarily mean that the particle decays freely.

4. Methods of obtaining information about the parities of particles

Conservation of parity is known not to hold in weak interactions. This fact was discovered experimentally by WU et al. (1957) after the possibility was pointed out by LEE and YANG (1956). Parity conservation has been tested in some strong interactions and found to hold, but tests have not been made in enough cases to give us complete confidence in the generality of this result. For example, parity conservation might be violated at extremely high energies.

If two particles are in a state of orbital angular momentum L, the parity of the orbital motion is $(-1)^L$. This follows from the property of the spherical harmonics under reflection

$$Y_L^M(\theta, \varphi) = (-1)^L Y_L^M(\pi - \theta, \varphi + \pi) . \tag{4.1}$$

If there are $n - 1$ conserved orbital angular momenta L_i associated with a state of n particles, the parity P of the state is given by

$$P = P'(-1)^{\Sigma L_i} , \tag{4.2}$$

where $(-1)^{\Sigma L_i}$ is the parity of the orbital motion and P' is the product of the intrinsic parities of the particles.

If a particle can be produced in a parity-conserving reaction in which no other particles are created or destroyed, the intrinsic parity of the produced particle is an observable. If, however, another particle is created or destroyed in the reaction, only the relative parity of the two particles can be measured. Many of the methods for determining intrinsic parities of particles are in reality methods of determining the relative parity of two or more particles. By the relative parity of several particles we mean the product of their intrinsic parities.

A. Production of particles

1. A method of determining the intrinsic parities of particles has been proposed by A. BOHR (1959). We shall consider BOHR's method only in the simplest case, that of a two-body initial state with a relative momentum p and a two-body final state with relative momentum q. Consider a reflection R with respect to the plane containing p and q. This reflection is equivalent to parity inversion followed by a rotation

of 180° about the normal $\hat{n}$ to the production plane. The eigenstates of R are eigenstates of the linear momentum operator rather than of angular momentum. It is apparent that the linear momenta do not change under R. Therefore, if the particles have no spin, the only thing that happens under R is that the initial and final wave functions get multiplied by a sign which depends on the intrinsic parities of the particles. Assuming parity is conserved, we have the selection rule $P = +$, where P is the product of the intrinsic parities of the particles. This same result can be obtained by noting that the orbital angular momentum must be the same in initial and final states if the particles have no spin.

If the particles have spin, let $\hat{n}$ be the quantization axis for the spin. The operator denoting a rotation about n is $e^{i\,\theta_n J_n}$ where J_n is the total spin component along $\hat{n}$ and θ_n is the angle of rotation. For a rotation of 180°, the spin wave function will be multiplied by a factor $e^{i\pi J_n} = (-1)^{J_n}$. Thus, conservation of parity implies that P is given by

$$P_i(-1)^{J_{\text{in}}} = P_f(-1)^{J_{\text{fn}}}, \tag{4.3}$$

where J_{in} is the component of the total spin along the n axis in the initial state and J_{fn} is the same for the final state. In general this relation is most useful when polarizations can be measured or when all the particles but one have spin 0.

2. A theorem relating the cross section for production of a particle in the forward direction to its parity has been given by LITHERLAND (1961). Applications have been given by many authors (CALDWELL 1961, MORPURGO 1963, BERMAN 1963, LICHTENBERG 1963b). The theorem states that in the reaction a + b → c + d, where a, b, and c have spin zero and d has spin J, then parity conservation implies that forward-backward production is forbidden if

$$P_a P_b P_c P_d = (-1)^{J+1}, \tag{4.4}$$

where P_a, P_b, P_c and P_d are the intrinsic parities of the particles. LITHERLAND proved the theorem by noting that if (4.4) holds, the Clebsch-Gordan coefficients multiplying the spherical harmonics Y^0_L in the scattering amplitude are zero for all L consistent with parity conservation. But since the Y^M_L vanish in the forward direction unless $M = 0$, the amplitude vanishes in the forward direction.

Since it may seem mysterious why the appropriate Clebsch-Gordan coefficients are zero, we give an alternative proof. We assume that forward production is possible and show that this implies $P_a P_b P_c P_d = (-1)^J$. Consider a state of two spinless particles e, f with the same total angular momentum, z-component, and parity as d. It is sufficient to prove the theorem for the process a + b → c + e + f. (One can imagine d decaying into e and f with parity conservation.) Now the parity of d is related to the product of the intrinsic parities of e and f as follows

$$P_d = P_e P_f (-1)^J \tag{4.5}$$

where the $(-1)^J$ comes from the parity of the orbital motion of e and f. Furthermore, if d is produced in the forward direction, the orbital wave function of e and f in their own rest frame must be the spherical harmonic

$Y_J^0(\theta, \varphi)$. This follows because with spinless particles in the initial state, the angular momentum can have no component in the beam direction. But Y_J^0 has a finite amplitude in the forward direction. Thus, if a + b → → c + e + f is allowed when e and f have *resultant* momentum in the forward direction, a possible configuration is one in which all particles are along a line.

Consider a reflection in a plane perpendicular to that line. Since the relative momenta of the particles are unchanged by the reflection, and since all particles have spin zero, the parity of the reflected state must equal the parity of the original state multiplied by $P_a P_b P_c P_e P_f$. Parity conservation implies that this product is $P_a P_b P_c P_e P_f = +1$. Otherwise the mirror reaction will be inequivalent to the original one. Combining with $P_d = P_e P_f (-1)^J$, we obtain $P_a P_b P_c P_d = (-1)^J$ as the condition for forward production to be allowed.

3. In the scattering of a spin 0 particle by a spin $1/2$ particle, the differential cross section is independent of the parity. This means that if the scattering amplitude in any partial wave corresponding to a given angular momentum and parity is replaced by an equal scattering amplitude with the same angular momentum and opposite parity, the cross section is unchanged. Thus, if there is a resonance in any angular momentum state J, the parity of the state cannot be determined from a measurement of the differential cross section alone. This was first noticed in pion-nucleon scattering by MINAMI (1954) and is known as the Minami ambiguity.

We give a proof due to DYSON and NAMBU which was outlined by BETHE and DE HOFFMANN (1955). If the wave function of the final state is ψ, the angular distribution is given by $I(\Theta, \Phi) = |\psi|^2$, where a sum over the spin states of the final-state spin $1/2$ particle is implied.

Now consider the operator $\boldsymbol{\sigma} \cdot \hat{\boldsymbol{\Theta}}$ where $\boldsymbol{\sigma}$ is the Pauli spin operator and $\hat{\boldsymbol{\Theta}}$ is a unit vector in the Θ direction. This operator is a scalar under rotation, and thus commutes with the total angular momentum. But $\boldsymbol{\sigma} \cdot \hat{\boldsymbol{\Theta}}$ is odd under the parity operation. Therefore the wave function

$$\psi' = \boldsymbol{\sigma} \cdot \hat{\boldsymbol{\Theta}} \, \psi \tag{4.6}$$

has parity opposite that of ψ. But the angular distribution of the state is given by

$$|\psi'|^2 = |\boldsymbol{\sigma} \cdot \hat{\boldsymbol{\Theta}} \, \psi|^2 = |\psi|^2 = I(\Theta, \Phi), \tag{4.7}$$

since $\boldsymbol{\sigma} \cdot \hat{\boldsymbol{\Theta}}$ is hermitian and $(\boldsymbol{\sigma} \cdot \hat{\boldsymbol{\Theta}})^2 = 1$. Thus, in order to obtain information about the parity, the polarization of the final-state spin $1/2$ particle must be measured.

B. Decays of particles

We give some theorems relating the parity of an unstable particle to its spin and to the parities of its decay products. It is assumed, of course, that parity is conserved in the decay.

1. If a particle of spin J decays into two spinless particles, its parity P is given by

$$P = P'(-1)^J, \qquad (4.8)$$

where P' is the product of the intrinsic parities of its decay products. The proof follows trivially from the fact that parity is a multiplicative quantum number. We made use of this result earlier in proving the Litherland theorem about forward production.

2. If a spinless particle decays into three spinless particles, its parity equals the product of the intrinsic parities of its decay products. The proof follows from noting that the two orbital angular momenta of the three final particles must be both even or both odd to make a state of angular momentum zero. Therefore the parity of the orbital motion is even.

3. If a particle decays into a spin $1/2$ particle and a spin 0 particle, the Minami ambiguity applies. Thus, no matter what the state of polarization of the decaying particle, the parity cannot be determined without a measurement of the polarization of the final spin $1/2$ particle.

Many proposals have been made to determine both the spin and parity from a measurement of the decay angular distribution $I(\theta, \varphi)$ and the polarization $\boldsymbol{P}(\theta, \varphi)$ of the final spin $1/2$ particle. We merely quote one such method, that given by Byers and Fenster (1963). From the measured values of $I(\theta, \varphi)$ and $\boldsymbol{P}(\theta, \varphi)$ we can form the quantities A_l^μ and B_l^μ defined by

$$A_l^\mu = \int I(\theta, \varphi)\, Y_l^\mu(\theta, \varphi)\, d\Omega, \qquad (4.9)$$

$$B_l^\mu = \left(\frac{l+1}{2l+1}\right)^{1/2} \sum_m P_m(\theta,\varphi)\, Y_{l-1}^{\mu-m}(\theta,\varphi)\, I(\theta,\varphi)\, (1, l-1, l; m, \mu-m) +$$
$$+ \left(\frac{l}{2l+1}\right)^{1/2} \sum_m P_m(\theta,\varphi)\, Y_{l+1}^{\mu-m}(\theta,\varphi)\, I(\theta,\varphi)\, (1, l+1, l; m, \mu-m), \qquad (4.10)$$

where the P_m are components of the polarization given by

$$P_0 = P_z, \quad P_{\pm 1} = \pm\, (P_x \pm i P_y)\, 2^{-1/2} \qquad (4.11)$$

and the coefficients $(1, l \pm 1, l; m, \mu - m)$ are Clebsch-Gordan coefficients*. The angular distribution is assumed to be normalized. Conservation of parity implies that A_l and B_l are zero for odd l. The spin J of the decaying particle can be determined if, for any $l > 0$ and any μ, the pair of numbers (A_l^μ, B_l^μ) is different from zero. The formula is

$$2J + 1 = \gamma\, [L(L+1)]^{1/2}\, B_l^\mu/A_l^\mu, \qquad (4.12)$$

where

$$\gamma = 1 \quad \text{if} \quad L = J - 1/2$$
$$\gamma = -1 \quad \text{if} \quad L = J + 1/2, \qquad (4.13)$$

L being the orbital angular momentum of the final state. The intrinsic parity P of the particle and the relative intrinsic parity P' of particles

---

* Our notation for a Clebsch-Gordan coefficient is $(j_1 j_2 j; m_1 m_2)$ with commas where ambiguity might otherwise arise. Some other notations are $(j_1 j_2 m_1 m_2 | j m)$ and $C(j_1 j_2 j | m_1 m_2)$. See Edmonds (1957), Rose (1957), Wigner (1959).

into which it decays are related by

$$P = P'(-1)^L . \tag{4.14}$$

4. If a particle decays into a spin $^3/_2$ particle and a spin 0 particle, there is no Minami ambiguity. Thus, there is at least a possibility of obtaining evidence concerning the parity from the decay angular distribution without measuring polarizations. Observing a decay into a spin $^3/_2$ particle is not as remote a possibility as one might think at first. This is because a number of the known excited baryon states have sufficiently large masses to decay into other excited baryons plus pions. A second possible reason is that high-mass states may have high spins. If so, it may be more favorable (by way of minimizing the orbital angular momentum and lowering the centrifugal barrier) for these particles to decay into $^3/_2$ (or higher) excited states rather than into their ground states.

The decay angular distribution of a particle of unknown polarization cannot be predicted uniquely. Furthermore, if the particle decays into a spin $^3/_2$ particle, the angular distribution cannot be predicted in general even if the polarization is known. This is because (assuming parity is conserved) there are two possible values of the orbital angular momentum consistent with a given spin and parity of the excited baryon. Thus, any statement about the relative parity depends on dynamical assumptions. It may be reasonable to assume that if the energy released in the decay is low, then the decay proceeds through the lower of the two allowed values of L. Then if the particle is polarized, the decay angular distribution gives information about the parity as well as the spin.

5. Adair analysis and generalizations

A. Selection of production events

Consider the decay of a particle which is produced in a reaction with an angular distribution $I(\Theta)$ and subsequently decays.

Since the particle will decay nonisotropically only if it is produced polarized or aligned, it may be advantageous to select production events for analysis so as to maximize the polarization or alignment. Any selection process not dependent on the decay will not produce a bias, but of course the decay angular distribution obtained by rejecting some events will have a larger statistical uncertainty than if all the data were used.

One way of selecting production events is to give each particle produced at an angle Θ a statistical weight $W(\Theta)$. Particularly useful weight functions are

$$W(\Theta) = \sin^2\Theta, \quad W(\Theta) = \cos^2\Theta, \quad W(\Theta) = \sin\Theta \cos\Theta , \tag{5.1}$$

which emphasize particles produced in the equatorial region, the polar region and the in-between region respectively.

B. Particles produced in the forward and backward directions

ADAIR (1955) has carried the idea of selecting production events to its logical conclusion. He has proposed a method which, for certain kinds of production processes, insures in principle that the particles of interest are produced completely aligned. ADAIR's method is to consider particles produced only in the forward and backward directions relative to the incident beam.

Consider the process

$$a + b \rightarrow c + d \,,$$

where a is a spin 0 particle (e.g. a pion or kaon) and b is either an un-polarized target particle of spin $^1/_2$ (e.g. a proton) or a particle of spin 0 (e.g. He⁴). The production amplitude A^m for the reaction in the forward direction may be written

$$A^m = \sum_{\mu} A_{\mu} \chi^{\mu}_{J_c} \chi^{m-\mu}_{J} \,, \tag{5.2}$$

where $\chi^{\mu}_{J_c}$ and $\chi^{m_d}_{J}$ $(m_d = m - \mu)$ are the spin wave functions of particles c and d respectively and the coefficients A_{μ} contain sums of spherical harmonics: $A_{\mu} = \sum \alpha_{L\mu} Y^0_L(\Theta = 0)$, where the angle-independent coefficients $\alpha_{L\mu}$ depend on the dynamics of the interaction. The magnetic quantum number m is the component of the target particle spin along the beam axis: if $J_b = 0$, then $m = 0$; if $J_b = ^1/_2$, then $m = \pm^1/_2$. The reason for the simplifying restrictions

$$A_{\mu} = \sum_{L} \alpha_{L\mu} Y^0_L(\Theta = 0) \,,$$

$$m_d = m - \mu \,, \tag{5.3}$$

is that the orbital angular momentum, not only in the inital state, but also in the final state for forwardly produced particles*, can have no component in the beam direction.

If particle c has non-zero spin, the analysis to determine the spin of particle d is quite complicated, although usually some information can be obtained. We restrict ourselves to the case in which $J_c = 0$. Then the expression for the amplitude A^m reduces to a single term:

$$A^m = A_0 \chi^m_J \,. \tag{5.4}$$

If $J_b = 0$, then $\chi^m_J = \chi^0_J$, and d is produced in a pure state. (A pure state with $m = 0$ is a completely aligned state.) If $J_b = ^1/_2$, then d is produced in a mixed state with equal amounts of the states $\chi^{1/_2}_J$ and $\chi^{-1/_2}_J$; again the spins of d are completely aligned. (If the target is polarized, or if parity is not conserved in the production process, the amounts of $\chi^{1/_2}_J$ and $\chi^{-1/_2}_J$ are not equal in general.)

Thus, for either a spin 0 target or for an unpolarized spin $^1/_2$ target, the spins of the produced particles are completely aligned, and we can obtain information about the spins from their decay angular distributions.

---

* In the following, our statements about forwardly produced particles also apply to particles which are produced in the backward direction.

C. Particles produced away from the forward direction

Since there is zero probability for particles to be produced exactly forward or backward, particles produced at non-zero angles must also be included in the analysis, and a criterion must be set up for how small the production angle must be to be acceptable. ADAIR (1955) has given such a criterion, but it depends on a dynamical assumption about the range of the interaction giving rise to the production.

It is a property of the spherical harmonics near $\Theta = 0$ that

$$|Y_L^1|/|Y_L^0| = \frac{1}{2} [L(L+1)]^{1/2} \Theta \approx L\Theta . \tag{5.5}$$

Thus, if L_{max} is the maximum orbital angular momentum entering the production process, then, near the forward direction, amplitudes other than the one containing Y_L^0 will in general be less than $L_{max}\Theta$ times the dominant amplitude. So if events are chosen only at angles Θ less than Θ_{max} where $L_{max}\Theta_{max} \lesssim 1$, the Adair analysis will be approximately correct. But L_{max} can be approximated by $kR = L_{max}$ where k is the momentum of either of the produced particles in the c.m. system and R is the range of the interaction. Thus, the criterion for an event to be useful is $\Theta < \Theta_{max} = 1/kR$. For a proton target, $R \approx 1/M_\pi$ where M_π is the pion mass. If a heavy nucleus is the target, R is its radius.

At an energy just above the threshold for production, we have $L_{max} = 0$, and all events are acceptable for the Adair analysis. This is obvious, since if the orbital angular momentum is zero, then so is its component along the beam axis. At the opposite extreme, in a coherent production process at high energy, we obtain the same result that all events are acceptable. This result, which is only approximate, can be seen as follows: For the process to be coherent, the maximum momentum transfer q to the nucleus must be of order $qR = 1$. But at high energy and at small angles $q = k\Theta$. Thus the cutoff angle in a coherent process is

$$\Theta = 1/kR \tag{5.6}$$

and is equal to the maximum useful angle for the Adair analysis.

The approximation is not so good in intermediate cases. As a specific example, assume a particle of spin $\frac{3}{2}$ and a particle of spin 0 are produced in a state with $L = 1$, $J = \frac{1}{2}$. The decay angular distribution of all spin-$\frac{3}{2}$ particles is then isotropic. The decay angular distribution of those particles produced at angles

$$\Theta \leq \Theta_{max} = 1/L = 1$$

is

$$I(\theta) = 1 + 0.8 \cos^2\theta .$$

This compares with the angular distribution

$$I(\theta) = 1 + 3 \cos^2\theta$$

expected if Θ is 0 or 180°.

The Adair analysis may be extended to apply to cases in which the production angular distribution is small or even zero in the forward direction (BERMAN 1963, MORPURGO 1963, LICHTENBERG 1963b).

For example, consider the production of integral-spin particles in a case where the parities of the particles are such that production in the forward direction is rigorously forbidden (Litherland theorem). Then the original Adair analysis fails. However, by making use of the property of spherical harmonics that

$$|Y_L^2|/|Y_L^1| = \frac{1}{4} [(L + 2)(L - 1)]^{1/2} \Theta \approx L\Theta \qquad (5.7)$$

near the forward direction, we can deduce that the spin wave functions χ_J of the decaying particle will be

$$\chi_J = \chi_J^1 + \chi_J^{-1}, \qquad (5.8)$$

rather then χ_J^0. Again information about the spin can be obtained from the decay angular distribution.

D. Decay angular distributions for integral spin particles

We assume that a particle with integral spin is produced in the forward direction, and that its spin wave function is given by χ_J^0. Suppose first that the particle decays into two spin 0 particles. In its own rest system its decay angular distribution $I(\theta)$ is given by

$$I(\theta) = |Y_J^0(\theta, \varphi)|^2, \qquad (5.9)$$

where $Y_J^0(\theta, \varphi)$ is an ordinary spherical harmonic, θ being measured with respect to the beam direction. For particles produced in the forward direction, the decay must be independent of φ, since no production plane is singled out. In Table 5.1(a) we give $I(\theta)$ for the first few values of J. Once the spin is known, the parity P of the particle is given in terms of the product of the intrinsic parities of the decay products P' by the equation

$$P = P'(-1)^J, \qquad (5.10)$$

provided parity is conserved in the decay.

If the particle decays into a spin one particle and a spin zero particle, its spin is uniquely predictable from its decay angular distribution only if (1) parity is conserved in the decay and (2) the spins and parities of the particles are related by Eq. (5.10). The angular distribution is given in terms of the spherical harmonics by

$$I(\theta) = |Y_J^1(\theta, \varphi)|^2 \qquad (5.11)$$

and is given in Table 5.1(b) for the first few values of J. The angular distribution is again independent of φ, since there is no production plane. If $P = P'(-1)^{J+1}$, the spin is not predictable from the decay angular distribution. This is because (assuming parity is conserved) the orbital angular momentum of the decay products is an unknown mixture of $L = J - 1$ and $L = J + 1$. However, if the energy of the decay products is not too high, it may be argued that the amount of $L = J + 1$ is small relative to $L = J - 1$ because the centrifugal barrier is significantly higher in the $L = J + 1$ state. If the contribution from $L = J + 1$

Table 5.1. Decay angular distribution $I(\theta)$ of a particle of integral spin J produced in the forward (or backward) direction with a spin 0 particle in a collision of two spinless particles. Here $x = \cos\theta$, where the decay angle θ is measured with respect to the beam direction

a) Decay into two spinless particles. If parity is conserved in the decay, then the product of the intrinsic parities of the decaying particle and the two final particles is $P = (-1)^J$

J	$I(\theta)$
0	1
1	x^2
2	$9x^4 - 6x^2 + 1$
3	$x^2(25x^4 - 16x^2 + 9)$

b) Decay with parity conservation into a spin one particle and a spin zero particle, if $P = (-1)^J$

J	$I(\theta)$
0	forbidden
1	$1 - x^2$
2	$(1 - x^2)\,x^2$
3	$(1 - x^2)(25x^4 - 10x^2 + 1)$

c) Decay with parity conservation into a spin one particle and a spin zero particle if $P = (-1)^{J+1}$. Since there are two possible values of the orbital angular momentum $(L = S \pm 1)$, $I(\theta)$ cannot be predicted uniquely without an additional assumption. Here it is assumed that only $L = S - 1$ contributes

J	$I(\theta)$
0	1
1	1
2	$3x^2 + 1$
3	$5x^4 - 2x^2 + 1$

Table 5.2. Decay angular distribution $I(\theta, \varphi)$ of a particle of integral spin J produced near the forward direction if the production cross section vanishes in the forward direction. Here, θ is measured with respect to the beam direction, φ is the azimuthal angle with respect to the production plane, and $x = \cos\theta$

a) Decay into two spinless particles

J	$I(\theta, \varphi)$
0	1
1	$(1 - x^2)\sin^2\varphi$
2	$x^2(1 - x^2)\sin^2\varphi$
3	$(5x^2 - 1)(1 - x^2)\sin^2\varphi$

b) Decay into a spin 1 particle and a spin 0 particle with $P = (-1)^J$

J	$I(\theta, \varphi)$
0	1
1	$1 - (1 - x^2)\sin^2\varphi$
2	$x^2 + (1 - 5x^2 + 4x^4)\sin^2\varphi$
3	$1 - 10x^2 + 25x^4 + (-1 + 131x^2 - 355x^4 + 225x^6)\sin^2\varphi$

c) Decay into a spin 1 particle and a spin 0 particle with $P = (-1)^{J+1}$, assuming $L = J - 1$

J	$I(\theta, \varphi)$
0	1
1	1
2	$x^2 + (1 - x^2)\sin^2\varphi$
3	$1 - 10x^2 + 25x^4 + (-40x^4 + 32x^2 + 8)\sin^2\varphi$

is negligible, the decay angular distribution and the spin are related as shown in Table 5.1 (c).

Even if the particle cannot be produced in the forward direction, the spin can still be predicted uniquely from the decay angular distribution, provided enough events can be obtained at small angles. If the decay is into two spinless particles, the angular distribution is given by

$$I(\theta, \varphi) = |Y_J^1(\theta, \varphi) + Y_J^{-1}(\theta, \varphi)|^2 , \qquad (5.12)$$

where φ is the azimuthal angle to the production plane. Some values of $I(\theta, \varphi)$ are given in Table 5.2(a). In Table 5.2(b) and 5.2(c) are given angular distributions for decay into a spin 1 and a spin 0 particle.

E. Decay angular distributions for half-odd-integral spin particles

Suppose a particle produced in the forward direction decays into a spin $^1/_2$ particle and a spin 0 particle. The decay angular distribution in the rest system of the produced particle is given by

$$I(\theta) = |Y_J^{1/_s}|^2 + |Y_J^{-1/_s}|^2 , \qquad (5.13)$$

where $Y_J^{1/_s}$ and $Y_J^{-1/_s}$ are angular momentum eigenfunctions obtained from combining spherical harmonics with spin-$1/_2$ wave functions. A sum over final spins is implied. Because of the Minami ambiguity, the angular distribution is independent of the parity. Furthermore, because parity is conserved in the production process and the target particles are unpolarized, $I(\theta)$ does not depend on whether parity is conserved in the decay. The angular distribution is given in Table 5.3(a) for the first few values of J.

Table 5.3. Decay angular distribution $I(\theta)$ of a particle of spin J produced in the forward direction with a spin 0 particle in a collision of a spin 0 particle with a spin $^1/_2$ particle. Here $x = \cos\theta$

a) Decay into a spin $^1/_2$ and a spin zero particle

J	$I(\theta)$
$^1/_2$	1
$^3/_2$	$1 + 3x^2$
$^5/_2$	$1 - 2x^2 + 5x^4$
$^7/_2$	$9 + 45x^2 - 165x^4 + 175x^6$

b) Decay into a spin $^3/_2$ particle and a spin 0 particle, assuming the orbital angular momentum is $L = J - 1/_2$ or $L = J - 3/_2$, depending on the parity. Here P is the product of the intrinsic parities of the unstable particle and those into which it decays

J	P	$I(\theta)$
$^3/_2$	$+$	1
$^3/_2$	$-$	$7 - 6x^2$
$^5/_2$	$-$	$1 + 2x^2$
$^5/_2$	$+$	$1 + 10x^2 - 10x^4$
$^7/_2$	$+$	$13 - 10x^2 + 45x^4$
$^7/_2$	$-$	$21 - 120x^2 + 465x^4 - 350x^6$

A measurement of the polarization of the final spin-$^1/_2$ particle resulting from the decay can yield information about the parity. However, if the target nucleons are unpolarized and parity is conserved in the production reaction, then particles produced in the forward direction are unpolarized. As a consequence, the spin-$^1/_2$ particles from their decay are also unpolarized. Thus, to determine the parity of particle which decays into a spin-$^1/_2$ particle and a spin-0 particle by means of an Adair analysis, the initial target nucleons must be polarized and the polarization of the final spin-$^1/_2$ particles must be measured. We list in Table 5.4 expected values of the polarization $P(\theta)$ multiplied by $I(\theta)$.

Table 5.4. Product of the decay angular distribution $I(\theta)$ and polarization $P(\theta)$ of a particle produced in the forward direction with a spin 0 particle in a collision of a spin 0 particle with a polarized spin $^1/_2$ particle. The decay is into a spin 0 particle and a spin $^1/_2$ particle, and P is the product of the intrinsic parities of the unstable particle and the particles into which it decays. Here $x = \cos\theta$

J	P	$I(\theta)\,P(\theta)$
$^1/_2$	$+$	1
$^1/_2$	$-$	$2x^2 - 1$
$^3/_2$	$-$	$5x^2 - 1$
$^3/_2$	$+$	$18x^4 - 15x^2 + 1$
$^5/_2$	$+$	$13x^4 - 10x^2 + 1$
$^5/_2$	$-$	$50x^6 - 65x^4 + 20x^2 - 1$

Suppose the produced particle decays with parity conservation into a spin-$^3/_2$ particle and a particle of spin 0. Then the possible values of the orbital angular momentum are a mixture of $L = J - ^3/_2$ and $L = J + ^1/_2$ if the spins and parities of the particles are related by

$$P = P'(-1)^{J+^1/_2},$$

where as usual P is the parity of the decaying particle and P' is the product of the parities of the final particles. Likewise, if the spins and parities are related by

$$P = P'(-1)^{J-^1/_2},$$

then the orbital angular momentum may be a mixture of $L = J - ^1/_2$ and $L = J + ^3/_2$. If the centrifugal barrier makes the contribution to the decay from the higher of the two allowed L values negligible, then both the spin and the parity can be uniquely predicted from the decay angular distribution (LICHTENBERG 1963a). The angular distributions for the first few values of the spin are given in Table 5.3(b).

F. Other extensions

BARSHAY (1963) has pointed out that any information about spins and parities which can be obtained from an Adair analysis with incident pions can also be obtained with linearly polarized photons. If the latter are used, however, the decay angular distributions and final polarizations must be measured with respect to the azimuthal angle φ as well as with respect to θ. If circularly polarized photons are used on a polarized target, a measurement with respect to φ is not needed, and indeed, yields no extra information. Unfortunately, it is not known at present how to obtain a sufficiently intense beam of circularly polarized photons.

Still another method of determining spins and parities of excited baryons has been proposed by C. ITZYKSON and M. JACOB (1963). These authors suggest looking at hyperon-antihyperon pairs produced in the forward direction in antiproton-proton annihilation and measuring polarization correlations among the final baryons. The method is quite complicated, and will not be discussed further here.

TREIMAN (1956) has applied the Adair argument to particle interactions at rest, in which an incident direction is not defined. The analysis still goes through if applied to the direction of the outgoing particles. TREIMAN (1962a) has also considered the additional information to be learned from an Adair analysis when parity is not conserved in the decay of one of the outgoing particles.

6. Dalitz plot and other plots

Consider either a production reaction or the decay of a particle leading to a final state consisting of three or more particles. If one has a sample of events of such a character, a question arises as to how to analyze the data to obtain useful information about the final state. One possibility is to calculate the invariant energy or mass* of two of the final state particles and to plot the number of pairs vs. the mass of a pair. Another possibility is to plot the angular distribution of one of the particles or the angular correlation between two of them. Both of these examples are plots of a frequency vs. a single variable of interest.

It is possible to obtain more information from the data by means of a plot in which two variables are given simultaneously. Each event is shown as a point on such a two dimensional plot, and information is obtained by observing the density distribution of points. For example, in a three-particle final state, the invariant mass of two of the particles may be plotted against the momentum transfer to the third.

See the review of ROSENFELD and HUMPHREY (1963) for a fuller discussion of the analysis of data, especially from bubble chamber pictures.

A. Dalitz plot

Of all point density plots, perhaps the most useful is the Dalitz plot (DALITZ 1953, FABRI 1954). Consider either a production reaction or a decay in which the final state consists of three particles. The transition

---

* By "invariant energy" or "mass" of a collection of particles we mean their total energy in their center-of-mass system. Sometimes the expressions "invariant mass" or "effective mass" are also used.

probability Γ for such a process is given by[*]

$$\Gamma = 2\pi \frac{|M|^2 D(E)}{\Pi_i 2E_i \Pi_f 2E_f}, \qquad (6.1)$$

where M is the invariant matrix element for the process, E is the total energy, $D(E)$ is the density of final states, and $\Pi_i 2E_i$ ($\Pi_f 2E_f$)· is the product of twice the energies of the initial (final) particles. The quantity $\varrho(E) = D(E)/(\Pi_f 2E_f)$ is relativistically invariant. For a 3-body final state it is given by

$$\varrho(E) = \frac{p_1^2 \, dp_1 p_2^2 \, dp_2 \, d\Omega_1 \, d\Omega_2}{8 (2\pi)^6 \, E_1 E_2 E_3 \, dE}, \qquad (6.2)$$

where the subscripts on the energies and momenta refer to the three final-state particles. In the c.m. system $\varrho(E)$ can be written in three convenient forms (after integrating over three angle variables)

$$\begin{aligned}
\varrho(E) &= k_a(E) \, dE_1 \, dE_2 \\
&= k_a(E) \, dT_1 \, dT_2 \\
&= k_b(E) \, dM_{12}^2 \, dM_{13}^2,
\end{aligned} \qquad (6.3)$$

where T_1 and T_2 are particle kinetic energies, M_{12}^2 and M_{13}^2 are the invariant masses of particles 1 and 2, and 1 and 3 respectively. Here k_a and k_b are functions of the total energy E but independent·of individual final state particle energies. This means that, for a given energy, E, the invariant phase space $\varrho(E)$ has a uniform density in the kinematically allowed region as a function of the variables (E_1, E_2), (T_1, T_2) or (M_{12}^2, M_{13}^2), where 1 and 2 refer to any two of the three particles. In a plot of an experimental distribution of three final-state particles in terms of any pair of these variables, the observed distribution of points gives useful information about the matrix element M.

B. Other plots

1. A four-body final state does not have the property of a 3-body state that the phase space has uniform density when plotted in convenient variables. Nevertheless, it is useful to plot the invariant mass of two of the particles against the invariant mass of the other two. The kinematically allowed region is a triangle in which the density of phase points decreases to zero at the edges. In a reaction producing a four-body final state at energies not far above threshold, it sometimes happens that the final state is consistent with the two step process

$$a + b \rightarrow a* + b*$$
$$a* \rightarrow c + d, \quad b* \rightarrow e + f,$$

where a* and b* are resonances. Evidence for such a process can be seen most easily on a triangular plot.

2. It is found that certain unstable particles are produced in some reactions at particular regions of momentum transfer, but not in other

---

[*] We set $\hbar = c = 1$.

regions. Thus, for example, in a three-body final state, it is useful to plot the invariant mass of two produced particles against the momentum transfer to the third. Such a plot is known as a Chew-Low plot.

3. To obtain information about the spin of a resonance, it is often useful to plot the angle of each decay against the momentum transfer in its production.

7. The nucleon

A. Quantum numbers of the nucleon

The mass of the proton is

$$M_p = 938.213 \pm 0.01 \text{ MeV}.$$

Its spin is $J = {}^1/_2$. The proton is stable, the experimental lower limit on its mean life being in excess of 10^{21} years. The stability of the proton is accounted for by the conservation of baryon number. As a consequence of this law, the least massive baryon has no other baryon to decay into; therefore it must be stable. The lightest baryon happens to be the proton. Of course the conservation of baryon number does not explain why the proton has any of its observed properties other than its stability.

The mass of the neutron is

$$M_n = 939.507 \pm 0.01 \text{ MeV}.$$

The spin of the neutron is also ${}^1/_2$. The neutron is unstable, decaying via the mode

$$n \rightarrow p + e^- + \bar{\nu}_e$$

with a mean life τ given by

$$\tau = 1013 \pm 26 \text{ sec}.$$

Protons and neutrons are fermions, as evidenced by the fact that the nuclei in which they are bound exhibit shell structure.

Because of the conservation of baryon number, a neutron cannot be created in a reaction unless another baryon is destroyed or an antibaryon created as well. For this reason, the intrinsic parity of the neutron can be measured only relatively to the parity of some other baryon. Therefore, we can arbitrarily define the parity of the neutron to be positive, and measure the parity of other baryons relative to that of the neutron. Similarly, because of the conservation of charge number, the parity of one charged particle can be defined to be positive. It is convenient to define the proton to be this charged particle.

It is sometimes said that the proton and neutron have the same parity. This is true by definition. The relative parity of proton and neutron cannot be measured without an additional assumption about the parity of some third particle. The selection rules which arise from the conservation of charge number and baryon number are called superselection rules. WICK, WIGHTMAN, and WIGNER (1952) have called attention to these rules and pointed out the ambiguities in parity and other quantum numbers which follow. Conservation of the two lepton numbers, if exact, lead to two additional superselection rules.

Evidence from proton-proton and neutron-proton scattering and from the energy levels of light nuclei shows that the interactions of protons and neutrons are charge independent. The proton and neutron are members of an isospin doublet, the nucleon, with the proton having $I_z = {}^1/_2$ and the neutron having $I_z = -{}^1/_2$. Since the proton has $Q = 1$ and the neutron $Q = 0$, the nucleon satisfies the Gell-Mann−Nishijima formula (Eq. 2.1) with hypercharge $Y = 1$.

The magnetic moment of the proton in nuclear Bohr magnetons, i.e. in units of $\mu_0 = e/2M_p$, is (WILSON 1963)

$$\mu = 2.79268 \pm 0.00001 .$$

Since the expected magnetic moment of a Dirac particle (a particle obeying the Dirac equation) is $\mu = 1$, the magnetic moment of the proton is often divided into two parts:

$$\mu_p = 1 + \varkappa_p$$

where the term 1 is called the Dirac moment and the term $\varkappa_p$ is called the anomalous or Pauli moment. Since the neutron has zero charge, it has only a Pauli moment given by

$$\mu_n = \varkappa_n = -1.9130 \pm 0.0001 .$$

B. Form factors of the nucleon

Experiments by HOFSTADTER (1956) and many others on the scattering of electrons by protons and neutrons (the latter bound in deuterons), have provided good evidence that protons and neutrons have structure. In other words, the observed scattering is quite different from the scattering that would be expected if the proton and neutron were point particles with their actual charges and magnetic moments. A measure of the departure of the observed scattering is given by the so-called form factors of the nucleon.

A good discussion of the theoretical aspects of electron-nucleon scattering has been given by DRELL and ZACHARIASEN (1961). A recent review of the experimental situation has been given by HAND, MILLER and WILSON (1963). Here we give only a brief summary of the facts.

The cross section for the scattering of electrons by nucleons has been calculated in lowest order perturbation theory in the electromagnetic interaction by ROSENBLUTH (1950). The Rosenbluth formula, in the approximation that the four-momentum transfer q is much greater than the electron mass is

$$\frac{d\sigma}{d\Omega} = \left(\frac{\alpha}{2E\sin\frac{\Theta}{2}}\right)^2 \frac{E'}{E}\left\{\cot^2\frac{\Theta}{2}\left[F_1^2 + \frac{q^2\varkappa}{4M_N^2}F_2^2\right] + \right.$$

$$\left. + \frac{2q^2}{4M_N^2}(F_1 + \varkappa F_2)^2\right\}, \tag{7.1}$$

$$q^2 = \frac{\left(2E\sin\frac{\Theta}{2}\right)}{1 + (2E/M_N)\sin^2\frac{\Theta}{2}}, \tag{7.2}$$

where $\alpha = {}^1/_{137}$ is the fine structure constant, E and E' are the incident and scattered electron energies, Θ is the scattering angle, and F_1 and F_2 are the so-called Dirac and Pauli form factors which are functions only of q^2. If the proton and neutron were point particles, F_1 and F_2 would be given by

$$F_{1p} = F_{2p} = 1, \quad F_{1n} = 0, \quad F_{2n} = 1. \tag{7.3}$$

The fact that F_1 and F_2 for both proton and neutron vary as q^2 varies, shows that these particles have structure. The alternative explanation, that quantum electrodynamics breaks down, is extremely unlikely, as quantum electrodynamics has been tested in experiments involving essentially only muons and electrons. It may be possible that part of the variation of the form factors arises from the breakdown of quantum electrodynamics, but some of the variation, at least, must arise from the effect of nucleon structure.

In the limit $q^2 \to 0$, the electron cannot distinguish between a point particle and a particle with structure, so that in this limit, the form factors approach the values given in Eq. (7.3).

Other form factors related to F_1 and F_2 are the electric and magnetic form factors G_E and G_M defined by

$$G_E(q^2) = F_1(q^2) - (q^2/4M_N^2)\varkappa F_2(q^2),$$
$$G_M(q^2) = F_1(q^2) + \varkappa F_2(q^2). \tag{7.4}$$

In terms of G_E and G_M, the Rosenbluth formula is

$$\frac{d\sigma}{d\Omega} = \left(\frac{\alpha}{2E\sin\frac{\Theta}{2}}\right)^2 \frac{E'}{E}\left\{\frac{\cot^2\frac{\Theta}{2}}{1 + q^2/4M_N^2}\left[G_E^2 + \frac{q^2}{4M_N^2}G_M^2\right] + \right.$$

$$\left. + 2\frac{q^2}{4M_N^2}G_M^2\right\}. \tag{7.5}$$

The form factors G_E and G_M are easier to obtain experimentally than F_1 and F_2. This is because in Eq. (7.1) there is a term containing the cross product F_1F_2, whereas in Eq. (7.5) there is no term containing $G_E G_M$.

SACHS (1962) has interpreted the Fourier transform of the charge and magnetic form factors as the spatial dependence of the charge and magnetic moment distributions respectively of the nucleon. However, since there seems to be no way to measure these distributions directly, this concept may have only pictorial usefulness.

The experimental values of G_E and G_M for the proton and neutron are given in Figures 7.1 through 7.4 from the review of HAND, MILLER and WILSON (1963). The form factors of the neutron have an uncertainty

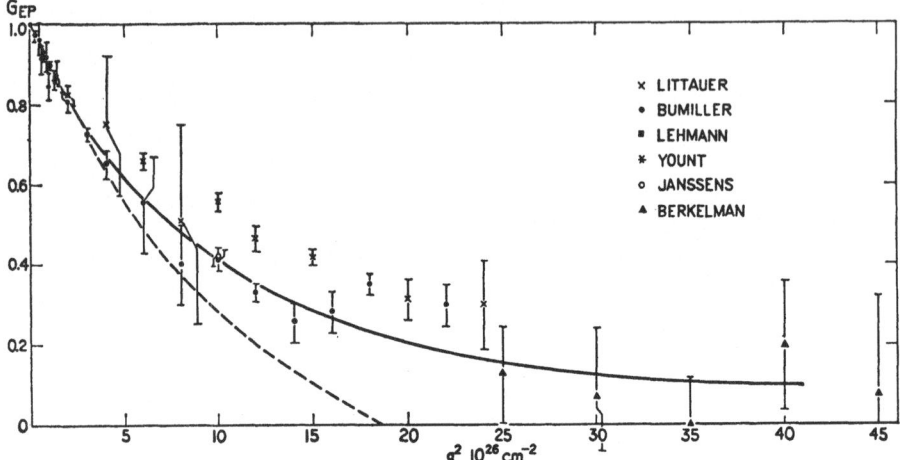

Fig. 7.1. The charge form factor of the proton $G_{E\,p}$ as a function of the square of the four-momentum transfer. The solid lines in Figs. 7.1 through 7.4 are semi-empirical fits to the data. The dashed lines are the predictions of a model in which the proton has a hard core. This model fails to fit the data (For further explanation see HAND 1963)

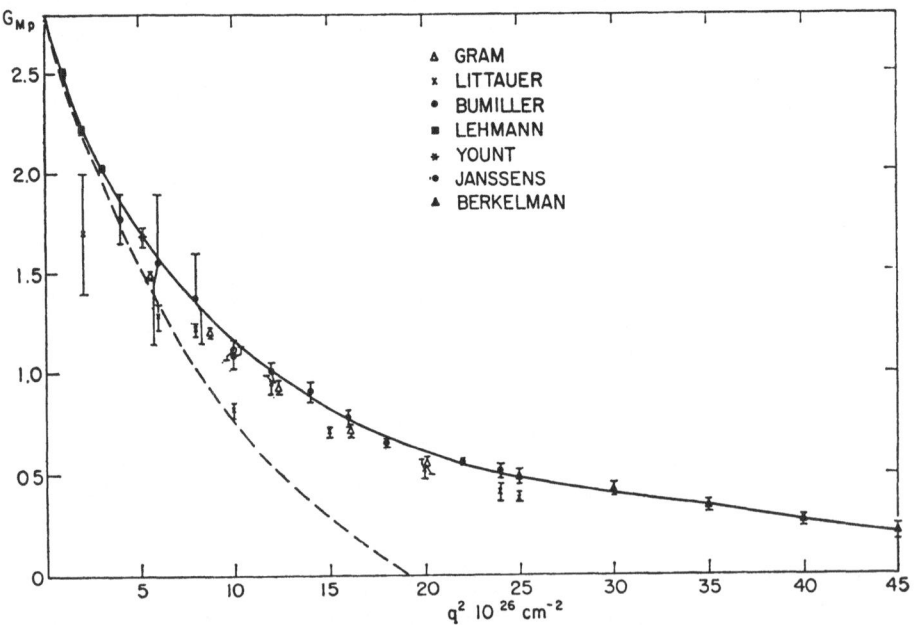

Fig. 7.2. The magnetic form factor of the proton $G_{M\,p}$ vs q^2 (HAND 1963)

in them arising from the fact that the neutron target in an experiment is bound in a deuteron. Thus, there are difficulties in the analysis arising from lack of knowledge of the structure of the deuteron.

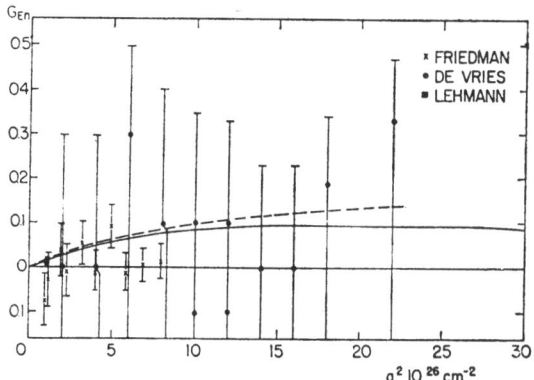

Fig. 7.3. The charge form factor of the neutron $G_{E n}$ vs q^2 (HAND 1963)

In Figures 7.1 through 7.4, the solid lines are semi-empirical fits to the data based on a model of photon coupling to vector mesons. The dashed lines are based on a model in which the nucleon has a hard core.

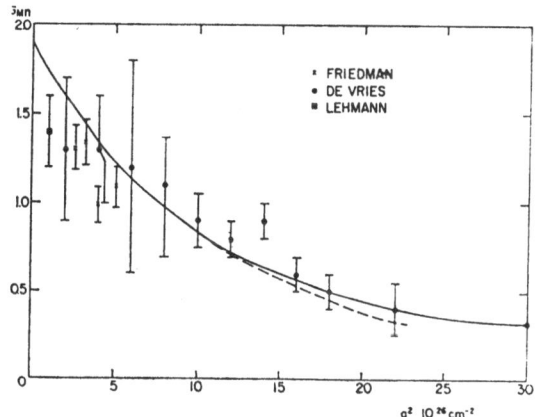

Fig. 7.4. The magnetic form factor of the neutron $G_{M n}$ vs q^2 (HAND 1963)

This model does not fit the data (HAND 1963). In Fig. 7.5 are shown the proton form factors including results at high momentum transfers (CHEN 1963).

To aid in interpreting the proton and neutron form factors, it is useful to introduce the isovector and isoscalar form factors G_V and G_S defined by

$$G_V = \frac{1}{2} (G_p - G_n)$$
$$G_S = \frac{1}{2} (G_p + G_n) . \tag{7.6}$$

To go from a proton to a neutron state requires a rotation of 180° in isospin space. The isovector form factor G_V changes sign (like a vector) under this rotation while the isoscalar form factor G_S remains invariant (like a scalar).

In lowest order perturbation theory, electron-nucleon scattering arises from the exchange of a virtual photon between the electron and nucleon. The electromagnetic interaction does not conserve isospin,

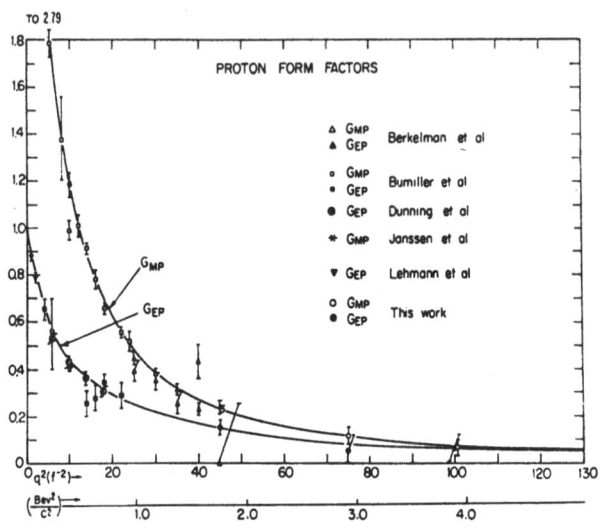

Fig. 7.5. Charge and magnetic form factors for the proton including results at high momentum transfer (CHEN 1963)

so that photon exchange contributes both to the isovector and isoscalar parts of the nucleon form factors. These form factors are thought to arise chiefly from the coupling of the photon to virtual vector mesons in a "cloud" around the nucleon.

Suppose the coupling is of the form $A_\mu V_\mu$ where A_μ is the electromagnetic field and V_μ is the field of a vector meson. If so, then the isovector form factor arises from coupling to vector mesons with $I = 1$, and the isoscalar form factor from coupling to mesons with $I = 0$. Since the electromagnetic interaction is invariant under charge conjugation, the mesons must have odd C parity. Then, to obtain the G parity of the mesons, we can use the relation $G = C(-1)^I$, Eq. (2.6), which holds for mesons of definite isospin and C parity. Thus, we find that a vector meson must have quantum numbers $I^G = 1^+$ or 0^- if its contribution to the nucleon form factor arises from single meson-photon coupling.

C. The deuteron

The deuteron is the only bound state of the proton and neutron. Its binding energy ε is

$$\varepsilon = 2.2245 \pm 0.0002 \text{ MeV} .$$

The total angular momentum of the deuteron is $J = 1$.

Since no bound state of two protons or two neutrons exists, the deuteron is an isosinglet. An $I = 0$ state composed of two particles with $I = 1/2$ is an antisymmetric state. Therefore the isospin wave function of the deuteron is antisymmetric under the interchange of the isospin coordinates of the two nucleons.

Since the wave function of the deuteron must be antisymmetric under the interchange of all the coordinates of the proton and neutron, the wave function must be symmetric under the simultaneous interchange of the space and spin coordinates of the two nucleons. Now the spatial wave function of a two-particle state depends only on the relative distance r between the two particles. Since $r = r_2 - r_1$, replacing r by $-r$ is equivalent to interchanging r_1 and r_2. Therefore, assuming the deuteron to have a definite parity*, its wave function must be either symmetric or antisymmetric under the interchange of its spatial coordinates.

To make the space-spin wave function symmetric, the spin part of the wave function must have the same symmetry as the spatial part. But there are only two possible spin wave functions of two nucleons, $S = 1$ which is symmetric and $S = 0$ which is antisymmetric. Since there can be no transitions between even and odd parity states, the total spin S of the two nucleons is a constant of the motion. The deuteron actually has spin $S = 1$, parity $P = +$. Its orbital angular momentum L is a mixture of $L = 0$ and $L = 2$, with $L = 0$ predominating. For the analysis leading to these conclusions, see BLATT and WEISSKOPF (1952).

A particle with $J = 1$ may have an electric quadrupole moment as well as a magnetic dipole moment. The quadrupole moment Q_d of the deuteron is (BLATT 1952, WILSON 1963)

$$Q_d = 2.82 \times 10^{-27} \text{ cm}^2 .$$

The magnetic moment of the deuteron in units of μ_0 is

$$\mu_d = 0.85741 \pm 0.00008 .$$

This differs by a small amount from the sum of the magnetic moments of the proton and neutron

$$\mu_p + \mu_n - \mu_d = 0.0222 .$$

8. The pion

A. Masses and lifetimes

The charged pion was discovered by LATTES, OCCHIALINI and POWELL (1947) in nuclear emulsions exposed to cosmic rays. Many of the properties of the pion have been discussed by BETHE and DE HOFFMANN (1955). Here we summarize a number of facts about the pion, including some information not contained in the book by BETHE and DE HOFFMANN.

* There is no reason to expect the deuteron (or any nucleus) to be an accidentally degenerate mixture of even and odd parity states.

Positive and negative pions have the same mass within experimental error. The mass of the positive pion is

$$M_{\pi^+} = 139.58 \pm 0.05 \text{ MeV}.$$

The lifetimes of the positive and negative pions are the same within experimental error. The mean life of the positive pion is

$$\tau_{\pi^+} = (2.547 \pm 0.027) \times 10^{-8} \text{ sec}.$$

This is a lifetime characteristic of a particle which decays by weak interactions. The best theoretical calculation of the pion lifetime was made by GOLDBERGER and TREIMAN (1958) using dispersion relations. They obtained agreement with experiment within 20%.

The observed decay modes of the positively charged pion are

$$\pi^+ \to \mu^+ + \nu_\mu$$
$$\to \mu^+ + \nu_\mu + \gamma$$
$$\to e^+ + \nu_e$$
$$\to \pi^0 + e^+ + \nu_e,$$

with the mode $\pi^+ \to \mu^+ + \nu_\mu$ being dominant. The experimental branching ratios into the rare decay modes are

$$R_1 = \Gamma(\pi^+ \to \mu^+ \nu \gamma)/\Gamma(\pi^+ \to \mu^+ \nu) = (2.8 \pm 1.2) \times 10^{-4} \qquad (8.1)$$
$$R_2 = \Gamma(\pi^+ \to e^+ \nu)/\Gamma(\pi^+ \to \mu^+ \nu) = (1.21 \pm 0.07) \times 10^{-4} \qquad (8.2)$$
$$R_3 = \Gamma(\pi^+ \to \pi^0 e^+ \nu)/\Gamma(\pi^+ \to \mu^+ \nu) = (1.15 \pm 0.22) \times 10^{-8}, \qquad (8.3)$$

where the Γ's are partial decay widths. A theoretical calculation of R_1 has been made using quantum electrodynamics and is in agreement with the experimental value. The ratio R_2 is in agreement with the prediction of the $V - A$ theory of weak interactions (FEYNMAN 1958, MARSHAK 1958, SAKURAI 1958). The ratio R_3 is in agreement with the prediction of the conserved vector current hypothesis (GERSTEIN 1955, FEYNMAN 1958). The corresponding rare decay modes of the π^- have not been looked for because a π^- coming to rest in nuclear matter is usually captured rather than decaying.

The neutral pion has the mass $M_{\pi^0} = 134,97 \pm 0,05$ MeV and decays electromagnetically into two photons. LINDENFELD et al. (1953) found that in 1.45% of the decays, one of the photons is internally converted to give a positron-electron pair called a Dalitz pair (DALITZ 1951). In $(0.73)^2$ % of the decays, both photons are converted. The calculation depends only on quantum electrodynamics and not on the details of the decay process. Theory and experiment are in reasonable agreement.

The lifetime of the π^0 has been obtained by measuring the distance traveled by π^0's before decaying. Only decays in which at least one of the photons is internally converted are useful for the measurement. The lifetime is

$$\tau(\pi^0) = (1.05 \pm 0.18) \times 10^{-16} \text{ sec}. \qquad (8.4)$$

This lifetime is about a factor 3 longer than the lifetime calculated in lowest order perturbation theory using pseudoscalar pion-nucleon coupling.

B. Isospin, hypercharge and G parity

The π^+, π^0, and π^- mesons are members of an isospin triplet; i.e. the pion has $I = 1$. A measure of the deviation from isospin conservation in pion interactions is given by the fractional difference between the masses of the charged and neutral pions

$$(M_{\pi^+} - M_{\pi^0})/M_{\pi^+} = 0.033 . \tag{8.5}$$

It follows from the Gell-Mann–Nishijima rule (Eq. 2.1) that the pion has hypercharge $Y = 0$. Therefore the pion has a definite G parity. The G parity of the pion is odd, as can be seen from the following argument:
It is convenient to write the pion wave functions as follows

$$\pi^+ = (\pi_x + i\pi_y)/\sqrt{2}$$
$$\pi^0 = \pi_z \tag{8.6}$$
$$\pi^- = (\pi_x - i\pi_y)/\sqrt{2} .$$

Since the quantities π_x, π_y, and π_z are the rectangular components of a vector in isospin space, they transform in the following way under a rotation of 180° about the y axis

$$R_y \pi_x = -\pi_x, R_y \pi_y = \pi_y, R_y \pi_z = -\pi_z .$$

Then

$$R_y \pi^+ = -\pi^-, R_y \pi^0 = -\pi^0, R_y \pi^- = -\pi^+ . \tag{8.7}$$

To complete the argument, we have to know how the pion behaves under C. The existence of the electromagnetic decay $\pi^0 \to 2\gamma$ shows that the π^0 has even C parity, since the C parity of a state consisting of n photons is $C = (-1)^n$, and C is conserved in electromagnetic interactions. Combining the relations $C\pi^0 = +\pi^0$, $R_y\pi^0 = -\pi^0$, we obtain $G\pi^0 = -\pi^0$. Since the π^- and π^+ are antiparticles of each other, we can write $C\pi^+ = \pm \pi^-$ where the sign multiplying the π^- wave function is arbitrary. With either sign, the charged pions are seen to be eigenstates of G, and we can choose the $+$ sign to make the charged pions have odd G parity as does the π^0. It follows that a state of n pions has G parity

$$G = (-1)^n . \tag{8.8}$$

C. Spin of the pion

Although pions have been produced in many different reactions at many energies, they have always been observed to decay isotropically. This is good evidence that they have spin zero.

Additional evidence has been obtained that the spin of the π^+ is zero by a measurement of the cross section σ_1 for the reaction

$$p + p \to d + \pi^+$$

and the cross section σ_2 for the inverse reaction

$$\pi^+ + d \to p + p .$$

According to the principle of detailed balance, if both cross sections are measured at the same energy, they must be related by

$$\frac{\sigma_1}{\sigma_2} = \frac{3\,(2\,J+1)}{2}\,\frac{k^2}{K^2}\,,$$

where J is the pion spin and k and K are the c.m. momenta of the pion and proton respectively. Good agreement with experiment is obtained for $J = 0$. For details and further references see BETHE and DE HOFFMAN (1955).

The π^- must have spin 0 since it is the antiparticle of the π^+. By charge independence the π^0 must also have spin 0. Note that the π^0 cannot have spin 1, since if it did, it could not decay into two photons (YANG 1950).

Since pions have integral spin, they ought to be bosons. There is fairly good evidence that this is so, since a number of symmetric two-pion states have been observed, whereas an antisymmetric state of two pions has not been seen. Further evidence that pions are bosons will be discussed in connection with K meson decay.

D. Parity of the pion

The pion is a pseudoscalar particle; i.e., it has odd intrinsic parity. This fact was inferred from observation of the reaction (PANOFSKY 1951, CHINOWSKY 1954)

$$\pi^- + d \to n + n$$

with stopped pions. The argument (FERRETTI 1946) depends on the fact that the pion is captured by the deuteron in a state of orbital angular momentum $L = 0$. (The evidence in favor of capture in a state with $L = 0$ will be given in Section E.)

Since the pion has spin 0 and the deuteron has spin $J = 1$, the total angular momentum of the initial state is $J_i = 1$. Also, since the deuteron has positive parity, the parity of the initial state is just the intrinsic parity P_π of the π^-.

The final state consists of two identical fermions in a state of total angular momentum $J = 1$. Therefore the wave function of the two neutrons must be antisymmetric. If the two neutrons are in a symmetric spin state ($S = 1$), then their space wave function must be antisymmetic. This implies that the orbital angular momentum l of the two neutrons is odd. Alternatively, if the neutrons are in an antisymmetric spin state ($S = 0$), then the space wave function must be symmetric and l must be even. But in this case, the total angular momentum $J_f = l$ of the final state is even and cannot be reached from the initial $J_i = 1$ state. Therefore l is odd and so is the parity P_f of the final state. Conservation of parity in the reaction implies that $P_\pi = -$.

It can be deduced from isospin conservation that the π^+ and π^0 also have negative parity. For example, if the π^+ and π^- had opposite parity, the cross sections for the reactions

$$\pi^+ + d \to p + p \tag{8.9}$$

$$\pi^- + d \to n + n \tag{8.10}$$

would not be equal, in general, as required by charge independence. This follows because if the π^+ had positive parity, reaction (8.9) could not go with the incident pion in an $L = 0$ state, whereas of course reaction (8.10) does go from this state. At low incident pion energy, in particular, the cross section for reaction (8.9) would then be smaller than that for reaction (8.10) (barring some dynamical accident), and would have a different energy dependence. Experiments involving many other πN reactions amply support the assumption that isospin is conserved.

Since the π^+ and π^- have the same parity, the parity of a state of π^+ and π^- with orbital angular momentum L must be $P = (-1)^L$. But, as we have previously noted, as well as being members of an isospin triplet, the π^+ and π^- are antiparticles of each other. Therefore the π^+ and its antiparticle satisfy the relation

$$P = (-1)^L$$

predicted from field theory to hold for a boson-antiboson pair. This relation must also hold for the π^0 and its antiparticle, since the π^0 is its own antiparticle. Also since pions are bosons, two π^0's must have L even and positive parity.

E. Capture of negative pions in hydrogen, deuterium, and helium

A theoretical argument that a stopping π^- is captured by liquid hydrogen in an $L = 0$ state has been given by DAY, SNOW and SUCHER (1959). The argument, which is similar for hydrogen and deuterium, goes as follows:

A stopping π^- is typically captured by a hydrogen atom in a Bohr orbit with principal quantum number $n \approx 15$. The π^- is bound tightly by atomic standards, and wanders like a neutral object in the liquid until it passes close to a neighboring proton. The strong electric field of the nearby proton causes a Stark effect which mixes states of different orbital angular momenta. Of these states, only the $L = 0$ state has an appreciable amplitude at the position of the proton, and therefore absorption takes place from this state.

Experiments of FIELDS et al. (1960) and DOEDE et al. (1963) have lent support to this picture. These authors measured the time required by a π^- to go from a velocity of 0.01 (in units of the velocity of light) to nuclear capture in liquid hydrogen. The measured cascade time $t_c(H_2)$ is computed from the fraction of pions which decay instead of being captured. The cascade time is

$$t_c(\mathrm{H_2}) = (2.2 \pm 0.6) \times 10^{-12} \sec.$$

This time is between one and two orders of magnitude shorter than the time calculated (WIGHTMAN 1949, 1950, BURHOP 1952, p. 168) for the pion to make radiative and Auger transitions to the $2p$ state ($n = 2$, $L = 1$). Thus, it appears that the pion is captured from a state with large n and $L = 0$ long before it has a chance to reach the $2p$ state*.

Capture of negative pions by liquid helium proceeds in a somewhat different manner. As before, the pion is captured in a Bohr orbit with a large principal quantum number. However, since the helium nucleus is doubly charged, the bound state of a π^- and a helium nucleus is not neutral. Therefore this state cannot wander arbitrarily close to another nucleus because of the Coulomb repulsion. Because of this fact, the Stark effect is not likely to be as important in mixing states of different orbital angular momenta.

The cascade time of negative pions in liquid helium has been measured by FETKOVICH and PEWITT (1963) and by BLOCK et al. (1963). The observed cascade time

$$t_c(\mathrm{He}) = (3.1 \pm 0.6) \times 10^{-10} \sec$$

is in rough agreement with the result expected if the π^- makes only radiative and Auger transitions after its initial capture in a Bohr orbit.

This result indicates that the pion in fact reaches the $2p$ state in helium. Capture from the $2p$ state probably competes with the radiative transition to the $1s$ state.

The knowledge of the cascade times of the π^- in hydrogen and helium can be applied with some modification to the capture of other negatively charged particles in these liquids. Applications will be given later in connection with capture of K^- mesons and antiprotons.

9. The Λ hyperon

The Λ hyperon** was probably discovered in cosmic rays (ROCHESTER 1947) at about the same time as the pion was first seen. FRANZINETTI and MORPURGO (1957) have given an account of the discovery of the Λ and other strange particles, and have discussed methods by which the properties of these particles were obtained. A more recent discussion of the strange particles has been given by ADAIR and FOWLER (1963).

---

* Previous to the paper by DAY, SNOW, and SUCHER, it was thought that the π^- reached the $2p$ state by making radiative transitions. The question was whether the pion was captured from the $2p$ state or made a transition to the $1s$ state before being captured. BRUECKNER, SERBER, and WATSON (1951) calculated that it is much more probable (by a factor ≈ 30) for a pion in the $2p$ state to undergo a radiative transition to the $1s$ state than to be captured in the $2p$ state.

** A hyperon is a baryon with hypercharge not equal to the hypercharge of the nucleon; i.e., a hyperon has $Y \neq 1$ (or non-zero strangeness). Any particle with strangeness $S \neq 0$ is called a strange particle.

A. Mass, isospin and hypercharge

The Λ is a neutral baryon of mass

$$M = 1115.38 \pm 0.10 \text{ MeV} .$$

A measurement of its magnetic moment has been made. In units of $^1/_2 \, e/M_p$ it is

$$\mu = -1.5 \pm 0.5 .$$

Since no corresponding charged baryon of the mass of the Λ exists, the Λ has isospin $I = 0$. Then, from the formula of GELL-MANN and NISHIJIMA the Λ has hypercharge $Y = 0$.

The pion and the nucleon form a system with baryon number $B = 1$ and hypercharge $Y = 1$. A Λ has $B = 1$, $Y = 0$. Therefore, if a Λ is to be produced in a pion-nucleon collision, it must be produced together with a particle or particles which have quantum numbers $B = 0$, $Y = 1$. No state consisting only of pions, nucleons, and their antiparticles has these quantum numbers. However, the K meson (to be discussed later) does have $B = 0$, $Y = 1$. Another state with these quantum numbers consists of a nucleon and an antihyperon. Thus, if a Λ is to be produced in a pion-nucleon collision which conserves hypercharge, it must be produced in association with either a kaon or an antihyperon. The phenomenon of associated production was suggested by PAIS (1952) to explain the strong production and weak decay of mesons and hyperons.

B. Decay modes

The lifetime of the Λ is

$$\tau = (2.57 \pm 0.30) \times 10^{-10} \text{ sec} .$$

Its observed decay modes and partial decay rates (in percent) are

$$
\begin{aligned}
\Lambda &\to p + \pi^- & (66 \pm 4)\% \\
&\to n + \pi^0 & (34 \pm 3)\% \\
&\to p + e^- + \nu .
\end{aligned}
\tag{9.1}
$$

The observed rate of β decay is about $^1/_{20}$ as large as the value calculated on the assumption that the coupling constant is the same as in the β decay of the neutron. Many of the weak decay rates of the other strange particles are comparably reduced.

It is interesting to consider the relation between the quantum numbers of the Λ and those of its decay products, excluding the quantum numbers of the leptons. We notice several facts about the decays of (9.1):

1. Hypercharge is not conserved but changes by $\Delta Y = +1$.

2. In the leptonic decay the charge number (of the strongly interacting particles) is not conserved, but changes by $\Delta Q = +1$.

3. Isospin is not conserved, but changes by $\Delta I = {}^1/_2$ or $^3/_2$. However, agreement with the observed branching ratio

$$\Gamma(\Lambda \to p\pi^-)/\Gamma(\Lambda \to n\pi^0) = 2$$

is obtained by assuming a final pion-nucleon state of $I = {}^1/_2$.

These facts are consistent with the following two selection rules for decays in which hypercharge is not conserved (GELL-MANN 1955).

1. In leptonic decays which do not conserve hypercharge $\Delta Y/\Delta Q = +1$ (or $\Delta S/\Delta Q = +1$).

2. In decays which do not conserve hypercharge, $|\Delta I| = {}^1/_2$.

Rule 2 implies rule 1 but not conversely. Rule 1 seems to hold in some weak decays of kaons and hyperons, but there is an indication of a violation in neutral K meson decays. Rule 2 seems to hold approximately in non-leptonic decays of kaons and hyperons, but seems to be violated in some leptonic decays. Since it is not our purpose to discuss the weak interactions in any detail, we shall forego pursuing this interesting subject. For further details and references see ADAIR and FOWLER (1963), JACKSON (1963), the Proceedings of the CERN Conference on High Energy Physics (1962) and the Proceedings of the Brookhaven Weak Interaction Conference (1963).

C. Spin and parity

It is observed that Λ hyperons produced in the forward direction in the reaction

$$\pi^- + p \rightarrow \Lambda + K^0$$

decay isotropically. From the Adair analysis, it is seen that the Λ has spin $J = {}^1/_2$. The decay angular distribution has also been measured using polarized Λ's. A polarized Λ decays into the mode $\Lambda \rightarrow p + \pi^-$ with an angular distribution

$$I(\theta) = 1 - \alpha P_\Lambda \cos\theta ,$$

where θ is the angle of the outgoing proton with the Λ (polarization P_Λ) and α is the asymmetry parameter. This parameter was measured by CRONIN and OVERSETH (1963)

$$\alpha = -0.62 \pm 0.07 .$$

Our sign convention differs from that of CRONIN and OVERSETH. The analysis of LEE and YANG (1958) shows that this large value of α is consistent only with $J = {}^1/_2$.

The fact that α is different from 0 shows that the final state of $\pi^- p$ contains both s and p waves. (An odd power of $\cos\theta$ arises from interference between even and odd L.) Therefore, the $\pi^- p$ state contains amplitudes of both even and odd parity. Thus, parity is not conserved in the decay.

The intrinsic parity of the Λ cannot be measured in a reaction which conserves hypercharge, since, as we have seen, conservation of hypercharge requires the Λ to be produced in association with another particle whose parity is unknown. Thus, it is only the relative parity of the two produced particles which can be measured in such a reaction.

If all interactions which do not conserve hypercharge also violate parity conservation, then the intrinsic parity of the Λ is not an observable. This is believed to be the case. It is customary to define the parity of the Λ to be even, and to measure the parities of other strange particles relative to that of the Λ.

D. Strong interactions with nucleons

The interaction of a Λ with a nucleon is of comparable strength to the nucleon-nucleon interaction. The force is a little weaker, however, since a bound state of a Λ and a single nucleon does not exist. (At least no such state has been observed.) The Λ can be bound in heavier nuclei, however. Such nuclei, discovered by DANYSZ and PNIEWSKI (1953), are called hyperfragments or hypernuclei. A list of hypernuclei and their binding energies is given by ADAIR and FOWLER (1963).

A number of authors, expecially DALITZ (1959), have analyzed the experimental information on the binding energies and lifetimes of hypernuclei. The main qualitative conclusions are that the Λ-nucleon force is spin dependent and that the interaction is stronger in the singlet spin state than in the triplet state. An important question bearing on the parity of the K meson is whether the hypernucleus $_\Lambda \mathrm{He}^4$ (a particle with two protons, a neutron and a Λ) has an excited state. Unfortunately, theory does not yield a definite answer, and a definite experiment has not been done.

10. The kaon

A. Mass, isospin and hypercharge

The kaon exists in two charge states, the K^+ and K^0, and thus has isospin $I = 1/2$. The masses of the K^+ and K^0 are

$$M_{K^+} = 493.98 \pm 0.14 \text{ MeV} \qquad M_{K^0} = 497.9 \pm 0.6 \text{ MeV} .$$

The mass difference is

$$M_{K^0} - M_{K^+} = 3.9 \text{ MeV} ,$$

$$(M_{K^0} - M_{K^+})/M_{K^+} = 0.0079 .$$

From the formula of GELL-MANN and NISHIJIMA, it is seen that the kaon has hypercharge $Y = 1$. This being so, an antikaon has $Y = -1$, and is distinct from a kaon. Therefore, there are two antikaons: $\overline{K}^0$ and $K^- = \overline{K}^+$. The interactions of the K and $\overline{K}$ with nucleons are very different, as the KN system has $Y = 2$, while the $\overline{K}N$ system has $Y = 0$.

In particular, there exist hyperons with $Y = 0$ but no known hyperons with $Y = 2$. Thus, the $\bar{K}N$ interaction often leads to hyperon production whereas (at least at low energies) the KN interaction does not.

B. Decays of the charged kaon

The K^+ decays weakly with a mean life

$$\tau = (1.227 \pm 0.008) \times 10^{-8} \text{ sec}.$$

The observed decay modes and the partial decay rates in per cent are (SHAKLEE 1964)

$$
\begin{aligned}
K^+ &\to \pi^+ + \pi^0 & (21.8 \pm 1.2)\,\% \\
&\to \pi^+ + \pi^- + \pi^+ & 5.0 \pm 0.2 \\
&\to \pi^+ + 2\pi^0 & 1.8 \pm 0.2 \\
&\to \mu^+ + \nu_\mu & 63.8 \pm 0.7 \\
&\to \mu^+ + \nu + \pi^0 & 3.0 \pm 1.0 \\
&\to e^+ + \nu + \mu^0 & 4.6 \pm 0.3 \\
&\to e^+ + \pi^+ + \pi^- + \nu & (2.3 \pm 0.7) \times 10^{-3}.
\end{aligned}
$$

There are also radiative decays similar to these except for an extra photon. The rare decay mode $K^+ \to e^+ + \pi^+ + \pi^- + \nu$ was seen by BISI et al. (1963) and by BIRGE et al. (1963). The branching ratio into this mode is from BIRGE et al., and agrees with theoretical estimates. No decays into the mode $K^+ \to e^- + \pi^+ + \pi^+ + \bar{\nu}$ have been seen. This latter decay is forbidden if the $\Delta Y = \Delta Q$ selection rule is valid.

The rate of decay of the K^+ into $\mu^+ + \nu$ is only about $1/20$ as large as the expected decay rate calculated on the assumption that the coupling constant leading to $\pi^+ \to \mu^+ + \nu$ and $K^+ \to \mu^+ + \nu$ are the same.

C. Spin

Kaons have been produced in many different reactions at many energies, but have always been observed to decay isotropically. This is good evidence that kaons have spin zero.

An analysis of the decay mode $K^+ \to \pi^+ + \pi^- + \pi^0$ has also provided some evidence in favor of this assignment. The method is to plot each event as a point on a Dalitz plot and to examine the distribution of points. The observed density is almost uniform within the kinematically allowed region, a fact that shows that the matrix element for the decay is almost constant. The simplest explanation for such an approximately constant matrix element is that the K has spin zero. The deviation from a constant distribution is probably due to final state interactions among the pions. If the K had higher spin, the decay pions would be emitted with orbital angular momenta. Therefore centrifugal barrier effects would tend to suppress low values of relative momenta of the pions. This, in turn, would lead to the depopulation of certain regions in the Dalitz plot. However, with the exception of spin 1, higher spins are not completely ruled out by this argument, as quite complicated matrix elements can occur which simulate a simple spin 0 matrix element. A further discussion of the use of the Dalitz plot to determine spins will be given in connection with the decays of the η and ω mesons.

D. Parity

Two of the observed decay modes of the K^+ are $K^+ \to \pi^+ + \pi^0$ and $K^+ \to \pi^+ + \pi^- + \pi^+$. Since a two-pion state with $J = 0$ has even parity and a three-pion state with $J = 0$ has odd parity, it follows that parity is not conserved in the weak decay*. Therefore, the parity of the K must be determined from its strong interactions.

Some evidence that the K^- has negative parity comes from the reaction

$$K^- + He^4 \to H^4 + \pi^0 . \tag{10.1}$$

The argument depends on a knowledge of the spin and parity of the hypernucleus $_\Lambda H^4$. This is thought to be $J^P = 0^+$ for the following reasons: The nucleus H^3 is known to have $J^P = 1/2^+$. Presumably, the Λ is bound in an $L = 0$ state, so the parity of $_\Lambda H^4$ is $P(_\Lambda H^4) = +$. (The parity of the Λ is defined to be positive.) With $L = 0$, the spin of $_\Lambda H^4$ must be either $J = 1$ or $J = 0$. But, as mentioned previously, an analysis of the data on hypernuclei leads to the conclusion that the Λ-nucleon force is more attractive in the spin 0 state than in the spin 1 state. If this is so, then the lowest energy state of $_\Lambda H^4$ has spin $J = 0$.

First suppose that no excited state of $_\Lambda H^4$ exists (none has yet been discovered). Then all the particles taking part in reaction (10.1) have spin 0. If this is so, the orbital angular momentum in the reaction, and therefore the parity of the orbital motion, must be the same in the initial and final state. Then, if parity is conserved, the product of the intrinsic parities of the particles must be even. This implies that the parity of the K^- must be odd (since the π has odd parity).

We also consider the remote possibility that $_\Lambda H^4$ has $J^P = 1^+$, still assuming that no excited state exists. If so, the evidence favors even parity for the K. The argument goes as follows: Stopped K^- mesons are observed to be captured in He^4, giving rise to reaction (10.1). The $_\Lambda H^4$ subsequently decays via the mode

$$_\Lambda H^4 \to He^4 + \pi^- \tag{10.2}$$

---

* It was the apparent existence of two mesons with the same mass and lifetime, one decaying into $\pi^+ \pi^0$ and the other into $\pi^+ \pi^- \pi^+$ that led LEE and YANG (1956) to suggest that both decays were different decay modes of the same particle and that parity was not conserved in weak interactions.

with an isotropic angular distribution with respect to the direction of flight of the $_\Lambda H^4$ (BLOCK 1962). The stopped K mesons are captured from Bohr orbits, presumably in s or p states. The evidence (KOPELMAN 1964) is that the K^-'s are captured primarily in p states, but we consider the possibility of both s- and p-state capture.

The problem is to obtain the angular correlation between two successive emissions of spin 0 particles (pions) from an unoriented nucleus (the $K^- - He^4$ bound state). The angular correlation depends on the initial angular momentum J (either 0 or 1 depending on whether the capture is from an s or p state), the angular momentum J' of the intermediate state ($_\Lambda H^4$) here assumed to be 1, and the angular momentum of the final nucleus (He^4) which is zero. The correlation also depends on the orbital angular momenta of the two pions and therefore on the parity of the system. From the theory of angular correlations (ROSE 1957), we obtain for the angular distribution $I(\theta)$ the results shown in Table 10.1. We see from the table that if the spin of $_\Lambda H^4$ is unity, only for p-state capture and positive K parity the angular distribution is isotropic, as required by experiment.

Table 10.1. Angular distributions of π^- mesons from the decay of $_\Lambda H^4$ hypernuclei produced in $K^- - He^4$ interactions at rest, assuming the spin of $_\Lambda H^4$ is 1. (Analysis indicates that the spin of $_\Lambda H^4$ is zero.) The angle θ is measured in the rest system of the $_\Lambda H^4$ with respect to the direction of its line of flight and L is the orbital angular momentum from which capture takes place

L	P_K	$I(\theta)$
0	—	forbidden
0	+	$\cos^2\theta$
1	—	$\sin^2\theta$
1	+	1 *

* This angular distribution is not necessarily isotropic, but is so under the simplest assumptions.

However, there is additional evidence that the parity of the K is in fact odd. This comes from combining two experiments to be discussed later. One of these (COURANT 1963) provides evidence that the Σ parity is even and the other (TRIPP '1962) that the relative $K - \Sigma$ parity is odd. If the results of these experiments are accepted, it follows that the spin of $_\Lambda H^4$ cannot be unity.

If reaction (10.1) proceeds via an excited state of $_\Lambda H^4$, which then decays by γ emission to the ground state, all the arguments connecting the parity of the K to the spin of $_\Lambda H^4$ break down. However, by far the most likely situation is that reaction (10.1) proceeds via the ground state of $_\Lambda H^4$, that the spin of this state is zero, and that the parity of the K is negative.

PAIS (1958) has suggested that the K^0 and K^+ might have opposite parity. It is hard to understand how this could be so if isospin is conserved in the strong interactions of K mesons.

E. Decays of the neutral kaon

In order to discuss the decay of the K^0, it is necessary to consider the decay of the $\overline{K}^0$ at the same time. Since the K^0 and $\overline{K}^0$ have opposite hypercharge, they are distinct. However, since they have $B = Q = 0$, they differ only in Y, which is not conserved in weak interactions. Therefore, the K^0 and the $\overline{K}^0$ can go into each other weakly.

However, it is believed that CP is conserved even in weak decays. Therefore it is useful to discuss the decays in terms of eigenstates of CP. We define the states K_1 and K_2 as follows

$$K_1 = (K^0 + CPK^0)\, 2^{-1/2}$$
$$K_2 = (K^0 - CPK^0)\, 2^{-1/2}. \tag{10.3}$$

From these definitions it follows that K_1 and K_2 are eigenstates of CP with eigenvalues plus and minus respectively. It is convenient to define the $\overline{K}^0$ as CPK^0 rather than CK^0, as is customary. Then

$$K_0 = (K_1 + K_2)\, 2^{-1/2}$$
$$\overline{K}_0 = (K_1 - K_2)\, 2^{-1/2}. \tag{10.4}$$

Now the neutral kaon is observed to decay into the modes $\pi^+ + \pi^-$ and $\pi^+ + \pi^- + \pi^0$. The first of these states is even under CP and the second may be either even or odd, depending on the orbital angular momentum of the pions. Therefore, if CP is conserved, the K_2 is forbidden to decay into two pions. Since the K_1 and K_2 are coupled differently to pions, there is no reason for their masses or lifetimes to be equal. In fact they are not. The beautiful prediction that the neutral kaon should be characterized by two distinct lifetimes was made by GELL-MANN and PAIS (1955a) using C invariance only, since at that time it was thought that C was conserved in weak interactions.

The observed lifetimes of the K_1 and K_2 are

$$\tau_{K_1} = (0.90 \pm 0.02) \times 10^{-10}\,\text{sec},$$

$$\tau_{K_2} = \left(6.3\,{}^{+1.0}_{-1.6}\right) \times 10^{-8}\,\text{sec}.$$

The lifetime of the K_2 is more than two orders of magnitude longer than that of the K_1 since it is forbidden to decay into the two-pion mode, with the largest available phase space.

The mass difference between the K_2 and K_1 is about

$$M_{K_2} - M_{K_1} \approx 1/\tau_{K_1} = 7 \times 10^{-6}\,\text{eV}$$

as measured by MULLER et al. (1960), BIRGE et al. (1960), and GOOD et al. (1961). The sign of the mass difference is based on a preliminary experiment of MEISNER et al. (1963) using a method suggested by CAMERINI, FRY and GAIDOS (1963).

The decay modes and partial decay rates of the K_2 are[*]

$$K_2 \rightarrow \pi^+ + \pi^- + \pi^0 \qquad (13 \pm 3)\,\%$$
$$\rightarrow 3\pi^0 \qquad 21 \pm 7$$
$$\rightarrow e^\pm + \nu + \pi^\mp \qquad 29 \pm 6$$
$$\rightarrow \mu^\pm + \nu + \pi^\mp \qquad 26 \pm 6\,.$$

The principal decay modes of the K_1 are

$$K_1 \rightarrow \pi^+ + \pi^- \qquad (69.4 \pm 1.0\%)$$
$$\rightarrow \pi^0 + \pi^0 \qquad 30.6 \pm 1.0\,.$$

Presumably, the K_1 also has most of the decay modes of the K_2. However, it is difficult to detect them as they represent less than 1% of the decays and can easily be confused with the decays of the K_2.

F. Statistics of the pion

The decays of the kaon provide evidence that pions are bosons. The first piece of evidence comes from a comparison of the lifetimes of the K^+ and K_1. The life of the K^+ is 200 times as long as the life of the K_1, even though both can decay into two pions. The inhibition of the decay of the K^+ is attributed to the fact that its decay into two pions violates the $|\Delta I| = \frac{1}{2}$ selection rule. This can be seen as follows: A state of $\pi^+ \pi^0$ with $J = 0$ must have $I = 2$ if the pions are bosons. Since the K^+ has $I = \frac{1}{2}$, the $\pi\pi$ decay satisfies the selection rule $|\Delta I| = \frac{3}{2}$ rather than $|\Delta I| = \frac{1}{2}$. However, if pions were not bosons, a state with $J = 0$ could have $I = 1$, and the $|\Delta I| = \frac{1}{2}$ rule could hold in K^+ decay. But then there would be no selection rule to inhibit the K^+ decay and its long life would be mysterious. This argument was given by GREENBERG and MESSIAH (1963) who attach little weight to it.

Another argument was given by HOLLADAY and VON BAEYER (1963). These authors point out that, if pions were not bosons, there could exist a $J = 0$ state of $\pi^+ \pi^-$ which is odd under CP. Then the K_2 could decay into it. Why it does not then would become another mystery.

11. The Σ hyperon

A. Masses, lifetimes, isospin, and hypercharge

The Σ exists in three charge states with masses

$$M_{\Sigma^+} = 1189.40 \pm 0.20 \text{ MeV}$$
$$M_{\Sigma^\bullet} = 1193.0 \pm 0.5$$
$$M_{\Sigma^-} = 1197.4 \pm 0.30$$

[*] See also LUERS et al. (1964) (Ed.).

and lifetimes

$$\tau_{\Sigma^+} = (0.78 \pm 0.03) \times 10^{-10} \text{ sec}$$
$$10^{-11} < \tau_{\Sigma^0} < 10^{-22}$$
$$\tau_{\Sigma^-} = (1.59 \pm 0.05) \times 10^{-10}\,.$$

The Σ^+ and Σ^- decay weakly into the modes (with partial decay rates in %)

$$
\begin{aligned}
\Sigma^+ &\to p + \pi^0 & (50.7 \pm 2.3)\% \\
&\to n + \pi^+ & 49.3 \pm 2.3 \\
&\to e^+ + \Lambda + \nu & \approx 10^{-2} \\
&\to \mu^+ + n + \nu & ? \\
\Sigma^- &\to n + \pi^- & \approx 100\% \\
&\to e^- + n + \bar{\nu} & 0.19 \pm 0.09 \\
&\to \mu^- + n + \bar{\nu} & 0.086 \pm 0.03 \\
&\to e^- + \Lambda + \bar{\nu} & \approx 10^{-2}\,.
\end{aligned}
$$

The leptonic decay rates of the Σ^+ and Σ^- are from WILLIS et al. (1963). One apparent Σ^+ decay into the mode $\Sigma^+ \to \mu^+ + \nu + n$ has been seen. This violates the $\Delta Y = \Delta Q$ selection rule. The other hypercharge-changing leptonic decay rates are only $1/20$ to $1/30$ as large as predicted by the universal Fermi interaction. The hypercharge-conserving decay rates $\Sigma^\pm \to$ $\to e^\pm + \Lambda + \nu$ are in agreement with the predictions of the Fermi theory.

The asymmetry parameters of the decays $\Sigma^+ \to n + \pi^+$, $\Sigma^+ \to p + \pi^0$, and $\Sigma^- \to n + \pi^-$ are

$$\alpha^+ < 0.04 \pm 0.11$$
$$\alpha^0 > 0.75 \pm 0.17$$
$$\alpha^- < 0.01 \pm 0.17\,,$$

where the superscripts on the α's give the charges of the pions emitted in the decays.

The amplitudes for the three Σ decays are related if the $|\Delta I| = 1/2$ selection rule is valid. Experiment indicates that the decays are consistent with the rule to a fairly good approximation. For further details see ADAIR and FOWLER (1963).

The Σ^0 decays electromagnetically into the modes

$$
\begin{aligned}
\Sigma^0 &\to \Lambda + \gamma \\
&\to \Lambda + e^+ + e^-.
\end{aligned}
$$

An accurate measurement of the fractional rate of decay into a Λ and a Dalitz pair has not been made. A theoretical estimate for this partial fractional rate is about $1/180$.

The hypercharge of the Σ is $Y = 0$, the same as that of the Λ. Thus, the stability of the Σ is not due to a new conservation law but to the "accident" that the $\Sigma - \Lambda$ mass difference is less than a pion mass. Otherwise the decay $\Sigma^0 \to \Lambda + \pi$ would be allowed to proceed via strong interactions.

B. Spin and parity

The spin of the Σ is observed to be $J = \frac{1}{2}$. The evidence comes from the Adair analysis and from the large value of the asymmetry parameter in the decay $\Sigma^+ \to p + \pi^0$.

Evidence that the parity of the Σ is positive comes from the decay $\Sigma^0 \to \Lambda + e^+ + e^-$. The method was suggested by FEINBERG (1958) and by FELDMAN and FULTON (1958) and the measurement was carried out by COURANT et al. (1963).

The invariant mass spectrum of the electron-positron pair depends on the $\Sigma - \Lambda$ relative parity (on the Σ parity assuming the Λ has even parity) and on the Dirac and Pauli form factors $F_1(q^2)$ and $F_2(q^2)$ of the $\Sigma - \Lambda$ electromagnetic transition. The four-momentum transfer q to the Λ in the transition is just the invariant mass of the $e^+ e^-$ pair which is limited to

$$q \leq M_{\Sigma_0} - M_\Lambda = \delta = 76.1 \text{ MeV}$$

or

$$q^2 \leq 0.15 \text{ Fermi}^{-2} .$$

From Figs. 7.1 through 7.4 we see that the nucleon form factors at $q^2 = 0.15$ have not changed much from their values at $q^2 = 0$. Unless the form factors for the $\Sigma - \Lambda$ transition are very different in character, this will also hold for them. However, since the Σ^0 and Λ are neutral, the Dirac form factor of the transition must vanish at $q^2 = 0$. Therefore, we introduce the form factor $f_1(q^2)$ defined by

$$F_1(q^2) = (q^2/M^2) \, f_1(q^2) , \tag{11.1}$$

where M is the mean mass of the Σ^0 and Λ, and make the approximation that

$$f_1(q^2) = f_1(0) , \quad F_2(q^2) = F_2(0) . \tag{11.2}$$

If this approximation is valid and if

$$f_1(0) \lesssim F_2(0) ,$$

then, in obtaining an expression for the mass distribution of the emitted electron-positron pair, the terms proportional to $f_1(0)$ may be neglected. In this approximation, the mass distribution $W(x)$ has the form

$$W(x) = C(x) (1 - x) \quad \text{for even parity}$$
$$W(x) = C(x) (1 + \tfrac{1}{2}x) \quad \text{for odd parity,} \tag{11.3}$$

where

$$x = q^2/\delta^2$$
$$C(x) = (1/x) (1 - x)^{1/2} (1 - x_0/x)^{1/2} (1 + x_0/2x)$$
$$x_0 = 4 M e^2/\delta^2 .$$

Note that the invariant mass of the electron-positron pair is just the four-momentum transfer.

In Fig. 11.1 are plotted the theoretical curves $W(x)/C(x)$ for both even and odd parity and the experimental points. The experiment clearly favors even parity. The full expression for the mass spectrum,

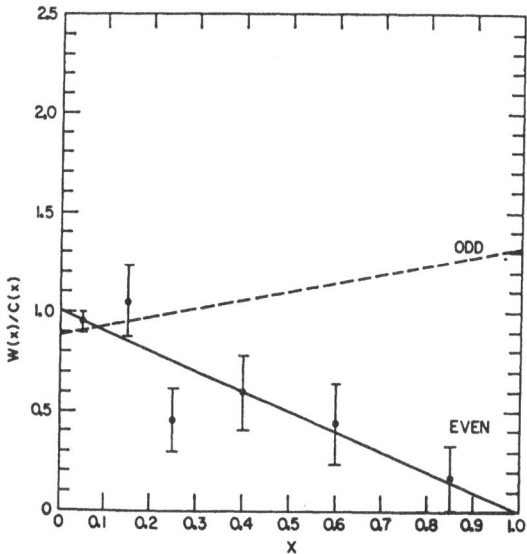

Fig. 11.1. The ratios of the number of events to the function $C(x)$, plotted against x, the square of the mass of the e^+e^- pair from the decay $\Sigma^0 \rightarrow \Lambda + e^+ + e^-$. The theoretical predictions for odd and even Σ parity are shown, assuming a negligible Dirac form factor for the transition. Spectra are normalized to the same number of events (COURANT 1963)

not neglecting f_1, has been given by EVANS (1962). From the expression of EVANS, it can be seen that if the ratio f_1/F_2 is about 10, the experiment does not distinguish between even and odd parity. However, there is no known mechanism which would lead to such a large Dirac form factor for the transition.

C. Strong interactions

The Σ cannot be bound in nuclei for an observable length of time because the strong interaction

$$\Sigma + N \rightarrow \Lambda + N \tag{11.4}$$

proceeds with an estimated lifetime of 10^{-23} sec. However, if the Σ-nucleon force is strong enough, there may occur bound states in which reaction (11.4) is forbidden because of the conservation of charge. For example, there might exist bound states of

$$\Sigma^+ p, \quad \Sigma^+ pp, \quad \Sigma^- n, \quad \Sigma^- nn.$$

None of these states has ever been seen, and they probably do not exist. The absence of these states does not imply that the Σ-nucleon force is weaker than the Λ-nucleon force, as the corresponding states Λp, Λpp etc., have not been seen either.

12. The Ξ hyperon

The Ξ hyperon is an isospin doublet (Ξ^0, Ξ^-) with hypercharge $Y = -1$. The cross sections for producing the Ξ in πN and $\overline{K}$N collisions are much smaller than the corresponding cross sections for producing the Λ or Σ. For this reason, not as much is known about the Ξ as about these other two hyperons.

The mass of the Ξ^- is

$$M = 1320.8 \pm 0.4 \text{ MeV} .$$

Its lifetime is (CONNOLLY 1963)

$$\tau_- = (1.74 \pm 0.18) \times 10^{-10} \text{ sec} .$$

Its observed decay modes and branching fractions are (CONNOLLY 1963)

$$\Xi^- \to \Lambda + \pi^- \quad \approx 100\%$$
$$\to \Lambda + e^- + \nu .$$

Parity is not conserved in the decay of the Ξ^-. The measured value of the asymmetry parameter in its decay is (CONNOLLY 1963)

$$\alpha_- = +0.47 \pm 0.16 .$$

The mass of the Ξ^0 is

$$M = 1316 \pm 3 \text{ MeV}, \quad M_{\Xi^-} - M_{\Xi^0} = 5 \pm 3 \text{ MeV} .$$

The Ξ^0 lifetime is

$$\tau_0 = \left(3.9 \begin{array}{c} +1.4 \\ -0.9 \end{array}\right) \times 10^{-10} \text{ sec} ,$$

and its only observed decay mode is

$$\Xi^0 \to \Lambda + \pi^0 .$$

The value of the asymmetry parameter is poorly known. One measured value (CONNOLLY 1963) is

$$\alpha_0 = 0.52 \pm 0.54 .$$

The large value of the asymmetry parameter in Ξ^- decay favors spin $J = 1/2$ for the Ξ, but the errors are too large to rule out $J = 3/2$. Too few events have been collected for the Adair analysis to give the spin.

The parity of the Ξ can in principal be determined, e.g., from the study of the reaction

$$K^- + p \to \Xi^- + K^+ .$$

However, thus far the parity of the Ξ has not been measured. It is usually assumed[*] that the Ξ, like the N, Λ, and Σ, has quantum numbers $J^P = 1/2^+$.

---

[*] The spin is now established to be $1/2$ (CARMONY 1964). Further results of these authors: $\tau_0 = (1.77 \pm 0.12) \, 10^{-10}$ sec, $\tau_0 = \left(3.5 \begin{array}{c} +1.0 \\ -0.8 \end{array}\right) 10^{-10}$ sec, $\alpha = 0.62 \pm 0.12$, $M_{\Xi^-} - M_{\Xi^0} = 6.1 \pm 1.6$ MeV (Ed.).

13. Selection rules

A. Meson decays

If a meson decays into strongly interacting particles with a full width at half maximum Γ which is about 1 MeV or greater, it is reasonable that it decays via strong interactions. If the width is too small to be measured by usual techniques but the particle travels too short a distance to be measured before decaying, the decay is probably either strong or electromagnetic. Finally if the particle has a lifetime of the order of 10^{-10} sec or longer, the decay is weak.

As we have remarked previously, the quantum numbers of a meson are Y, I, I_z (or alternatively Q), J, P, G (if $Y = 0$) and C (if $Y = Q = 0$). We assume that in decays via strong interactions all these quantum numbers are separately conserved, in electromagnetic interactions I and G may not be conserved, and in weak interactions only J, Q and CP are necessarily conserved. These conservation laws lead to selection rules in the decays of mesons into other mesons and photons. It is usually assumed that time reversal invariance also holds in all interactions, but since T is antiunitary, it leads to no selection rules.

We can obtain information about the quantum numbers of a meson by looking at its decay rate, at its branching ratio into its various modes, and at the quantum numbers of its decay products. We consider possible decays into pions, photons, K mesons, and other mesons. To get the selection rules, we list the quantum numbers of some of these multiparticle states. These quantum numbers have been discussed previously, but are collected here for convenience. In the following, n refers to the number of particles of a given kind and the quantum numbers refer to the total quantum numbers of the multiparticle state.

An n-pion state has the following quantum numbers

$$n\pi : Y = 0, \quad G = (-1)^n$$
$$n\pi^0 : C = + . \tag{13.1}$$

A two-pion state has the following relations among its quantum numbers

$$\pi^{\pm}\pi^0 : P = (-1)^J, \quad (-1)^J = (-1)^I$$
$$\pi^+\pi^- : C = P = (-1)^J, \quad (-1)^J = (-1)^I, CP = + \tag{13.2}$$
$$2\pi^0 : J \text{ even}, I \text{ even}, P = C = + .$$

These relations are a special case of a general relation discussed previously in which a meson of spin J and its antiparticle in a state of orbital angular momentum L and total spin S have the following quantum numbers

$$C = (-1)^{L+S}, \quad P = (-1)^L, \quad CP = (-1)^S . \tag{13.3}$$

Some relations among the quantum numbers of a three-pion state are

$$3\pi : J = 0 \quad \text{implies} \quad P = - . \tag{13.4}$$

In addition

$$\pi^+\pi^-\pi^0 : I = 0 \quad \text{implies} \quad C = - ,$$
$$3\pi^0 : C = +, I \neq 0 . \tag{13.5}$$

There are the following restrictions on states consisting of pions and photons

$$\pi\gamma : J \neq 0$$
$$\pi^0\gamma : J \neq 0, C = -$$
$$2\gamma : J \neq 1, C = + \tag{13.6}$$
$$n\pi^0 2\gamma : C = +$$
$$\gamma n\pi^0 : C = - .$$

Next consider a state of $K\overline{K}$. The following relations hold among the quantum numbers

$$K\overline{K} : P = (-1)^J, \quad G = (-1)^{J+I}$$
$$K^0\overline{K}^0, K^+K^- : C = P = (-1)^J, \quad CP = + . \tag{13.7}$$

It is conceivable that a meson might decay into the modes $K_1 K_1$, $K_1 K_2$, or $K_2 K_2$ without going through an intermediate $K^0\overline{K}^0$ state. Neither K_1 nor K_2 is an eigenstate of C, P or Y, but the states $K_1 K_1$, $K_1 K_2$, and $K_2 K_2$ have the following relations among their quantum numbers

$$K_1 K_1 \text{ and } K_2 K_2 : CP = +, J \text{ even}$$
$$K_1 K_2 : CP = (-1)^{J+1} . \tag{13.8}$$

These relations follow from the assumption that K_1 and K_2 are eigenstates of CP with eigenvalues plus and minus respectively. Comparing these relations with the result that $CP = +$ for $K^0\overline{K}^0$, we see that any state that can decay into $K_1 K_1$ or $K_2 K_2$ can also decay into $K^0\overline{K}^0$. On the other hand, a state which decays into $K_1 K_2$ can only decay into $K^0\overline{K}^0$ if J is odd.

Lastly we consider states of two non-identical mesons a and b

$$ab : G = G_a G_b, \quad P = P_a P_b(-1)^L$$
$$a^0 b^0 : C = C_a C_b = G_a G_b(-1)^{I_a + I_b} . \tag{13.9}$$

Tables 13.1 and 13.2 give the selection rules for meson decays into pions and photons. Some of these rules have been given previously (e.g., LICHTENBERG 1962, HENLEY 1962). Table 13.3 gives selection rules for meson decay into two mesons, one of which is not a pion. Table 13.4 gives selection rules for decay into two K mesons. In these tables a decay permitted by strong interactions is indicated by "yes", and a forbidden decay is indicated by the symbol for a conservation law which would be violated by the decay. This designation is not necessarily unique. For example in Table 13.1 we have indicated that a $J = 1$ meson is forbidden to decay into $2\pi^0$ mesons, stating that the occurence of such a decay would indicate that the pions did not obey Bose statistics. However, the decay could occur even if the pions are bosons if J were not conserved. We do not mean to imply that we expect any of these things to happen. Also sometimes more than one conservation law

Table 13.1. Selection rules for the decay into pions and photons of neutral mesons with hypercharge $Y = 0$. The quantum numbers J, P, I, and G apply to all members of a meson multiplet, but C applies only to the neutral member. A decay allowed by strong interactions is indicated by "yes", and forbidden decays are indicated by the symbol for a conservation law which would be violated by the decay. This designation is not necessarily unique, as the law might be conserved at the expense of violating another conservation law. A decay in which a photon appears is indicated as violating I if otherwise allowed. A decay which is forbidden by Bose statistics is indicated by st

Quantum numbers of meson I^G J^P C			Decay modes											
			$\pi^+\pi^-$	$2\pi^0$	2γ	$\pi^0\gamma$	$\pi^+\pi^-\pi^0$	$\pi^0 2\gamma$	$3\pi^0$	$\gamma 2\pi^0$	$\pi^+\pi^-\gamma$	$4\pi^0$	$\pi^+\pi^-2\pi^0$	$2\pi^+2\pi^-$
0^+	0^+	$+$	yes	yes	I	J	P	I	P	C	I	yes	yes	yes
0^-	0^+	$-$	CP	CP	C	J	P	C	P	I	I	C	G	G
1^+	0^+	$-$	CP	CP	C	J	P	C	P	I	I	C	yes	yes
1^-	0^+	$+$	G	G	I	J	P	I	P	C	I	G	G	G
0^+	0^-	$+$	CP	CP	I	J	G	I	G	C	I	yes	yes	yes
0^-	0^-	$-$	P	P	C	J	yes	C	C	I	I	C	G	G'
1^+	0^-	$-$	P	P	C	J	G	C	C	I	I	C	yes	yes
1^-	0^-	$+$	CP	CP	I	J	yes	I	yes	C	I	G	G	G
0^+	1^-	$+$	CP	st	J	C	G	I	G	C	I	yes	yes	yes
0^-	1^-	$-$	G	st	J	I	yes	C	C	I	I	C	G	G
1^+	1^-	$-$	yes	st	J	I	G	C	C	I	I	C	yes	yes
1^-	1^-	$+$	CP	st	J	C	yes	I	yes	C	I	G	G	G
0^+	1^+	$+$	P	st	J	C	G	I	G	C	I	yes	yes	yes
0^-	1^+	$-$	CP	st	J	I	yes	C	C	I	I	C	G	G
1^+	1^+	$-$	CP	st	J	I	G	C	C	I	I	C	yes	yes
1^-	1^+	$+$	P	st	J	C	yes	I	yes	C	I	G	G	G
0^+	2^+	$+$	yes	yes	I	C	G	I	G	C	I	yes	yes	yes
0^-	2^+	$-$	CP	CP	C	I	yes	C	C	I	I	C	G	G
1^+	2^+	$-$	CP	CP	C	I	G	C	C	I	I	C	yes	yes
1^-	2^+	$+$	G	G	I	C	yes	I	yes	C	I	G	G	G
0^+	2^-	$+$	CP	CP	I	C	G	I	G	C	I	yes	yes	yes
0^-	2^-	$-$	P	P	C	I	yes	C	C	I	I	C	G	G
1^+	2^-	$-$	P	P	C	I	G	C	C	I	I	C	yes	yes
1^-	2^-	$+$	CP	CP	I	C	yes	I	yes	C	I	G	G	G

Table 13.2. Selection rules for the decay of positively charged $I = 1$ mesons. A decay allowed by strong interactions is indicated by "yes" and a forbidden decay by the symbol for the conservation law which would be violated in the decay. For further explanation see the caption for Table 13.1

Quantum numbers of meson J^{PG}	$\pi^+\pi^0$	$\pi^+\gamma$	$\pi^+2\pi^0$ $\pi^-2\pi^+$	$\pi^+\pi^0\gamma$ $\pi^+2\gamma$	$\pi^+3\pi^0$ $\pi^-\pi^02\pi^+$
0^{++}	I	J	P	I	yes
0^{+-}	I, G	J	P	I	G
0^{-+}	P	J	G	I	yes
0^{--}	P	J	yes	I	G
1^{-+}	yes	I	G	I	yes
1^{--}	G	I	yes	I	G
1^{++}	P	I	G	I	yes
1^{+-}	P	I	yes	I	G
2^{++}	I	I	G	I	yes
2^{+-}	I	I	yes	I	G
2^{-+}	P	I	G	I	yes
2^{--}	P	I	yes	I	G

Table 13.3. Selection rules for the decay of $Y = 0$ mesons with $I \leq 1$ and other quantum numbers J^{PG} into a pion plus a meson (not a pion) of isospin $i \leq 2$ and other quantum numbers $j^{p i}$. An allowed decay is indicated by "yes" and a forbidden decay by the symbol of a conservation law which would be violated if the decay took place. In addition to the selection rules of the table, if $I = 0$ the decay is forbidden by I invariance unless $i = 1$. Also the decay (neutral meson → neutral meson $+ \pi^0$) is forbidden by C invariance unless $(-1)^{I+i}G_g = 1$. For further explanation see caption for Table 13.1

$j^{p i}$	0^{++}	0^{+-}	0^{-+}	0^{--}	1^{-+}	1^{--}	1^{++}	1^{+-}
J^{PG}								
0^{++}	P	P	G	yes	P	P	G	yes
0^{+-}	P	P	yes	G	P	P	yes	G
0^{-+}	G	yes	P	P	G	yes	P	P
0^{--}	yes	G	P	P	yes	G	P	P
1^{-+}	P	P	G	yes	G	yes	G	yes
1^{--}	P	P	yes	G	yes	G	yes	G
1^{++}	G	yes	P	P	G	yes	G	yes
1^{+-}	yes	G	P	P	yes	G	yes	G

Table 13.4. Selection rules for the decay of neutral mesons with $I \leq 1$ and $Y = 0$ into two K mesons. See caption for Table 13.1. By "yes" in the columns $K_1 K_1$ and $K_1 K_2$, we mean that the decay is allowed by strong interactions to proceed through the mode $K^0 \overline{K}^0$, with the $K^0 \overline{K}^0$ subsequently decaying weakly into $K_1 K_1$ or $K_1 K_2$. A forbidden decay is indicated by the symbol for a conservation law which would be violated in the decay. The symbol st means that the decay would violate Bose statistics

J^P	C	$K^+ K^-$	$K^0 \overline{K}^0$	$K_1 K_1$	$K_1 K_2$
0^+	$+$	yes	yes	yes	CP
0^+	$-$	CP	CP	CP	Y
0^-	$+$	CP	CP	CP	Y
0^-	$-$	P	P	P	CP
1^-	$+$	CP	CP	st	CP
1^-	$-$	yes	yes	st	yes
1^+	$+$	P	P	st	P
1^+	$-$	CP	CP	st	CP
2^+	$+$	yes	yes	yes	CP
2^+	$-$	CP	CP	CP	Y
2^-	$+$	CP	CP	CP	Y
2^-	$-$	P	P	P	CP

might be violated. For example we have indicated that a $J^P = 0^-$, $C = -$ meson is forbidden to decay into $\pi^+ \pi^-$ by P invariance. However, C would also not be conserved by such a decay. Finally, it is well to point out again that the selection rules make use of results from field theory, which have not been verified in every case.

B. Nucleon-antinucleon annihilation

We have seen that a state of a baryon and its antiparticle have the following relations among its quantum numbers

$$C = (-1)^{L+S}, \quad P = (-1)^{L+1}, \quad CP = (-1)^{S+1} . \tag{13.10}$$

It follows that for a single neutral meson to have the same quantum numbers as a possible bound state of a spin $1/2$ baryon and its anti-particle, the meson must have the following restrictions on its spin J, parity P, and C parity

If $J = 0$, then $C = +$,

If $P = (-1)^J$, then $C = (-1)^J$, $\qquad\qquad\qquad$ (13.11)

If $P = (-1)^{J+1}$, and $J > 0$, then no restriction on C.

The fact that the π^0 has $J = 0$, $C = +$ shows that it has the same quantum numbers as a possible bound state of a nucleon and antinucleon. If a meson has $I = 1$ and its neutral member can exist as a bound nucleon-antinucleon state, than so can its charged members, since $G = C(-1)^I$ for the meson and also for the $\overline{N}N$ pair.

In Table 13.5 are given the quantum numbers of possible bound states of a nucleon-antinucleon pair, with orbital angular momentum L restricted to $L \leq 1$. The selection rules for the production of various

Table 13.5. Quantum numbers of nucleon-antinucleon pairs in states of orbital angular momentum $L \leq 1$ and spin S. The selection rules for annihilation of these states into mesons and photons are the same as for decays of mesons with the same values of I^g and J^P. These selection rules are given in Tables 13.1 through 13.4. If the nucleon-antinucleon state has $I = 1$, the value of C refers to the neutral member of the triplet

I^G	J^P	C	L	S
0^+	0^+	$+$	1	1
1^-	0^+	$+$	1	1
0^+	0^-	$+$	0	0
1^-	0^-	$+$	0	0
0^-	1^-	$-$	0	1
1^+	1^-	$-$	0	1
0^+	1^+	$+$	1	1
1^-	1^+	$+$	1	1
0^-	1^+	$-$	1	0
1^+	1^+	$-$	1	0
0^+	2^+	$+$	1	1
1^-	2^+	$+$	1	1

mesons in the annihilation are the same as in the decay of mesons with the same quantum numbers as the pair. Thus, with the use of Table 13.5, Tables 13.1−13.4 include selection rules for nucleon-antinucleon annihilation in s and p states. Some of these selection rules have been given previously (LEE 1956a).

The argument of DAY, SNOW and SUCHER (1959) that stopped negative mesons are captured in s states in liquid hydrogen should apply to kaons and antiprotons as well. D'ESPAGNAT (1961) has proposed a test of s-state antiproton-proton annihilation based on the reactions

$$\overline{p} + p \rightarrow K^0 + K^0$$
$$\rightarrow K^+ + K^-. \qquad\qquad (13.12)$$

From Table 13.5 we see that the singlet and triplet s states of $\bar{\text{p}}$p have quantum numbers $J^{PC} = 0^{-+}$ (singlet) and $J^{PC} = 1^{--}$ (triplet). From Table 13.4 we see that reactions (13.12) can occur only from the state with $J^{PC} = 1^{--}$. We also see that the $\text{K}^0 \bar{\text{K}}^0$ state can go only into $\text{K}_1 \text{K}_2$, the final state $\text{K}_1 \text{K}_1$ being forbidden by Bose statistics. On the other hand, if the annihilation took place from a p state, the tables show that $\text{K}_1 \text{K}_2$ would be forbidden.

In an experiment to look for $\bar{\text{p}}$p annihilations into $\text{K}_1 \text{K}_1$ and $\text{K}_1 \text{K}_2$, ARMENTEROS et al. (1962) saw only $\text{K}_1 \text{K}_2$ pairs. This confirms that annihilation at rest takes place from the s state.

It is interesting to turn the argument around and assume s-state annihilation. It follows that if the $\bar{\text{p}}$p pair obeys the rules $C = (-1)^{L+S}$, $P = (-1)^{L+1}$ appropriate to fermions, then there exists a state of $\text{K}\bar{\text{K}}$ with the quantum number $C = (-1)^L$. The fact that the reaction

$$\bar{\text{p}} + \text{p} \rightarrow \text{K}^0 + \bar{\text{K}}^0 \rightarrow \text{K}_1 + \text{K}_1$$

is forbidden does not by itself show that kaons are bosons, however, since the reaction is forbidden by CP invariance as well as by Bose symmetry.

C. A parity

In addition to the quantum numbers we have discussed, there are perhaps still others which hold approximately in meson decay. The slight evidence for this consists in the fact that the decay rates of mesons into their various decay modes are not even approximately predictable from the usual conservation laws and phase space considerations. At present, we cannot say whether the discrepancies between predicted and observed decay rates are the results of complicated dynamical effects or whether there exist additional approximate selection rules in the decays.

One other possible selection rule might result from the approximate conservation of a quantum number R (FEINBERG 1959, GELL-MANN 1962, OKUBO 1963) or a related quantum number A (PEASLEE 1960, BRONZAN 1963).

The operator R, known as the R-conjugation or hypercharge reflection operator, replaces a state with charge number Q and hypercharge Y by a corresponding state with quantum numbers $-Q, -Y$. It is seen that all particles with $Q = Y = 0$ are eigenstates of R. Also, R acting on a K meson produces just an antikaon. However, R acting on a nucleon state produces a state with the mass and other quantum numbers of a nucleon but with $Y = -1$. No such state has ever been observed. However, in the approximation that the Ξ has the same mass and form factors as the nucleon, R acting on a nucleon state is a Ξ state (provided of course that the Ξ has $J^P = \frac{1}{2}^+$). This is probably not a good approximation since

$$M_\Xi - M_N = 380 \text{ MeV} .$$

The operator A is defined as RC, where C is the charge conjugation operator. All mesons and photons are eigenstates of A. However, since mesons are coupled strongly to baryons, conservation of A must be violated in the strong interactions of mesons. BRONZAN and LOW (1963) suggest that

$$A\pi^0 = -\pi^0 .$$

If this is so, the decay $\pi^0 \to 2\gamma$ is inhibited. The observed lifetime of the π^0 is about a factor 3 longer than the lifetime calculated by perturbation theory (BETHE 1955) and a factor 2 longer than that calculated by GOLDBERGER and TREIMAN (1958a). On the other hand, BRONZAN and LOW estimate a lifetime only $\frac{1}{40}$ of the observed life, and suggest that A conservation is good to $\approx 2\%$. This quantum number will be discussed further in connection with the decays of the η, ρ, ω, and φ mesons.

14. The ρ meson

A. Mass and width

The ρ meson was observed by ERWIN et al. (1961) as a correlation in the invariant mass spectrum of two final-state pions from the reactions

$$\pi^- + p \to \pi^+ + \pi^- + n$$
$$\to \pi^- + \pi^0 + p .$$

The two-pion mass spectrum is shown in Fig. 14.1. Peaks appear in both the $\pi^+\pi^-$ and $\pi^-\pi^0$ mass spectra, but the data are combined for better statistics.

What is plotted in Fig. 14.1 is not a histogram of the events but a rectangular ideogram. In such an ideogram, each event is given an equal weight (an equal rectangular area in the plot) but the width of the rectangle is proportional to the measurement error for the particular event. One may get the impression from the ideogram in Fig. 14.1, that there are many peaks in the two-pion mass spectrum. Most of these peaks are probably statistical fluctuations. To obtain a good estimate of the mass and width of a peak, a Gaussian ideogram is sometimes plotted.

In Fig. 14.1, the peak in the di-pion mass spectrum appears at a mass $M = 765$ MeV. The full width at half maximum Γ is more than 100 MeV. Since Γ is so large, there can be a large interference between pions from ρ decay and from background events. For this reason, the mass and width change from experiment to experiment. Average values are

$$M = 760 \text{ MeV}, \qquad \Gamma = 120 \text{ MeV} .$$

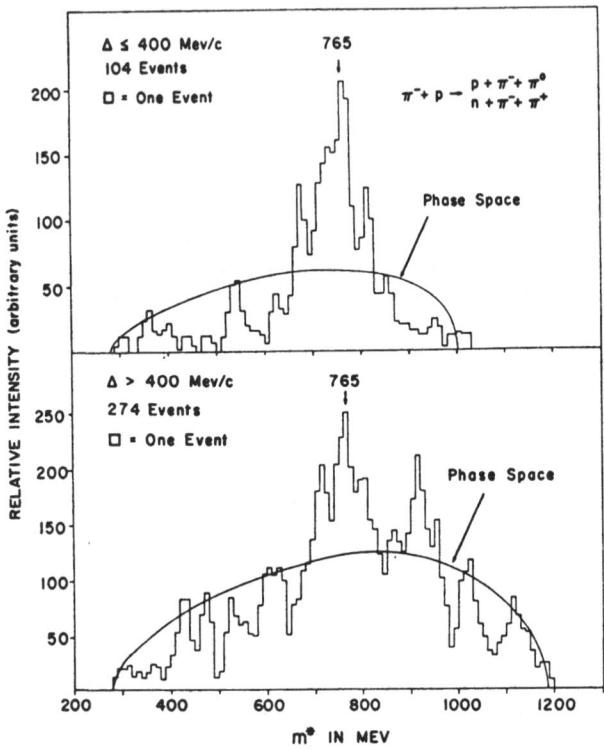

Fig. 14.1. The combined mass spectrum for the $\pi^- \pi^0$ and $\pi^- \pi^+$ system. The smooth curve is phase space as modified for the included momentum transfer and normalized to the number of events plotted. Events used in the upper distribution are not contained in the lower distribution (ERWIN 1961)

B. Isospin and G parity

Since the ρ decays strongly into two pions, it has positive G parity. (The G parity of n pions is $G = (-1)^n$.)

The fact that the ρ has been seen in three charge states shows that $I \geq 1$. Since a doubly charged ρ has not been seen (STONEHILL 1961) in $\pi^+ + p \to \pi^+ + \pi^+ + n$, whereas the ρ is very prominent peak in $\pi^+ + p \to \pi^+ + \pi^0 + p$, the ρ must have $I = 1$. If $I = 2$, the ratio R of the two reactions would be

$$R = (\pi^+ \pi^+ n)/(\pi^+ \pi^0 p) = 4$$

instead of the observed $R = 0$ within experimental error.

C. Spin and parity

The large decay width of the ρ makes it reasonable to assume that it decays via strong interactions and that isospin is conserved in the decay. This tells us that J is odd, as follows: A two-pion state with $I = 1$ has an isospin wave function which is antisymmetric under the interchange of the two pions. Since the pions obey Bose statistics, the two-pion space wave function must be antisymmetric to make the total wave function including isospin be symmetric. This implies odd orbital angular momentum for the two pions and therefore odd spin for the ρ. Since the parity of a two-pion system is $P = (-1)^J$, the parity of the ρ is odd.

The ρ is produced primarily in peripheral or low momentum-transfer collisions, as can be seen from Fig. 14.1. In this figure, the ρ is a prominent peak in the two-pion mass spectrum arising from pion-nucleon collisions with momentum transfers less than 400 MeV. However, in events with momentum transfers greater than 400 MeV, the figure shows that ρ production decreases. Other evidence that the ρ is produced with low momentum transfer comes from an experiment of ALFF et al. (1962). Fig. 14.2(a) shows that the production angular distribution of ρ events is peaked in the forward direction. (Small angles mean small momentum transfers.)

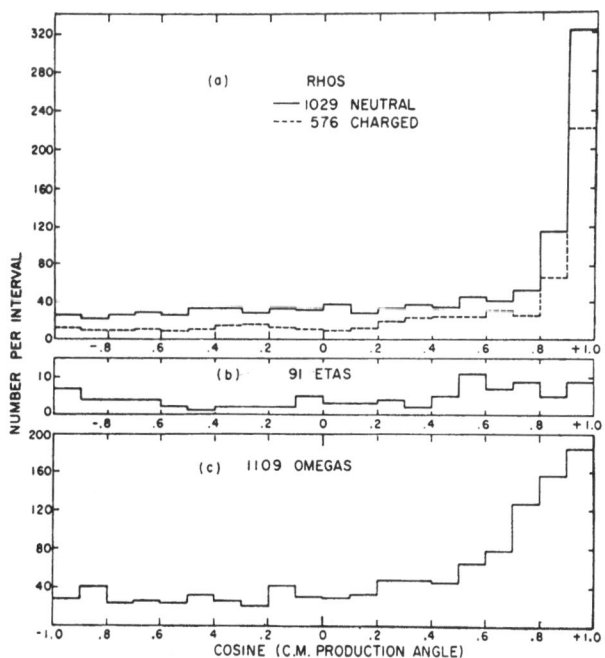

Fig. 14.2. Center-of-mass production angles for (a) rhos, (b) etas, and (c) omegas (ALFF 1962)

Because the ρ is produced primarily in the forward direction, an Adair analysis is useful. However, it does not yield a unique result. In fact, there are the following possibilities for the decay angular distribution $I(\theta)$ for a $J = 1$ ρ meson produced in the forward direction in the reaction $\pi + N \to \rho + N$:

$I(\theta) = \cos^2\theta$, nucleon spin not flipped,

$I(\theta) = \sin^2\theta$, nucleon spin flipped.

There cannot be interference between the spin-flip and non-spin-flip amplitudes, but they can add to give an isotropic distribution. The decay angular distribution of the charged ρ actually observed experimentally is

$$I(\theta) \approx \cos^2\theta .$$

Furthermore, although the ρ has been seen in many experiments, a $\cos^4\theta$ term has never been observed. This is good evidence for $J = 1$.

However, the decay angular distribution of neutral ρ mesons produced in $\pi^- p$ collisions contains odd powers of $\cos\theta$. The π^- from the ρ decay tends to go forward in the c.m. system. This is evidence that the ρ does not decay as a free particle of spin one.

With the quantum numbers $I^G = 1^+$, $J^P = 1^-$, the ρ has just the quantum numbers to contribute to the isovector form factor of the nucleon. FRAZER and FULCO (1959) predicted that such a meson (or pion-pion resonance) might exist.

D. One-pion-exchange model

The fact that the ρ is produced at low momentum transfers suggests the validity of a model in which the incident pion collides with a pion in the nucleon cloud, knocking it free. In this model, the ρ is a resonance in the pion-pion scattering cross section. The distinction between a particle and a dynamical resonance in this case may be just one of terminology.

A calculation of the one-pion-exchange diagram for the production of a spin one particle in the reaction $\pi^- + p \to \rho + N$ may be made. The result is that if the ρ has spin one, it decays with an angular distribution $I(\theta) = \cos^2\theta$, where θ is the angle the outgoing pion makes with the incident pion in the rest system of the ρ. Qualitatively, this can be seen as follows: In the rest system of the ρ, the incident pion and the virtual pion from the nucleon cloud are in a straight line. The orbital angular momentum of the two pions can have no component along this direction. Therefore, in this reference frame the wave function of the two pions after the collision is $Y_J^0(\theta)$, and the angular distribution is $I(\theta) = (Y_J^0(\theta))^2$. For $J = 1$, $I(\theta) = \cos^2\theta$. It is seen that the one-pion-exchange mechanism leads to the same result as an Adair analysis with a spin zero target particle. However, the angular distribution is measured with respect to an angle which is slightly different from the Adair angle.

Instead of considering only the interaction of the incident pion and the cloud pion to produce a ρ, one can consider the pion-pion interaction in other angular momentum states as well. The CHEW-LOW (1959) formula relating the cross section for pion production in πN collisions to the $\pi\pi$ total cross section is

$$\frac{d^2\sigma}{dq^2 dM^2} = \frac{f^2}{2\pi} \frac{q^2 M}{M_\pi^2} \frac{(\tfrac{1}{4}M^2 - M_\pi^2)^{1/2}}{(q^2 + M_\pi^2)^2 p^2} \sigma_{\pi\pi}(M) ,$$

where q is the four-momentum transfer to the nucleon, p is the incident pion momentum in the laboratory system, M is the invariant mass of the two final state pions and $\sigma_{\pi\pi}(M)$ is the pion-pion collison cross section at a center of mass energy M. This formula applies to the reactions $\pi^{\pm} + p \to \pi^{\pm} + \pi^0 + p$ occuring via one-pion exchange mechanism. For the formula to apply to $\pi^- + p \to \pi^+ + \pi^- + n$, the right hand side must be multiplied by a factor 2.

Strictly speaking, this formula is supposed to be valid in the limit $q^2 \to -M_\pi^2$. However, in practice, the formula has been assumed to hold in the physical region where q^2 is positive. With this assumption, the experimental data yield a peak value for $\sigma_{\pi\pi}(M)$ at $M \approx 750$ MeV, with

$$\sigma_{\pi\pi}(750 \text{ MeV}) \approx 12\pi\lambda^2 .$$

This value of the cross section is the unitarity limit for a $J = 1$ resonance, a result which favors $J = 1$. Note that the maximum value of $\sigma_{\pi\pi}$ is at a slightly lower energy than the peak in the production cross section.

Pion-pion scattering should also occur in angular momentum states other than $J = 1$. The best evidence that scattering does occur in other states comes from the asymmetric decay of the ρ^0. This asymmetry or odd power of $\cos\theta$ appearing in the decay may be due to interference with background events originating from processes other than one-pion exchange. However, it is more likely that the asymmetry is inherent in the pion-pion system. Such an odd power can arise only from interference between states of even and odd angular momenta.

This suggests that there may exist a large s- or d-state amplitude which interferes with the resonant p-wave amplitude[*]. The data do not exclude the possibility of an s-wave pion-pion resonance near 750 MeV.

Part of the structure in the peak may arise from interference between ρ^0 decay and the two-pion decay of the ω meson (to be discussed later). However, the ω is produced at higher momentum transfers than the ρ, so that this source of interference can be largely eliminated. Also since the ω has spin one, $\rho - \omega$ interference, if it occurs, cannot account for the forward-backward asymmetry in ρ^0 decay.

E. Treiman-Yang test

Part of the evidence for the ρ having spin 1 depends on the validity of the one-pion exchange model for ρ production at small momentum transfers. A test of the validity of the model has been proposed by TREIMAN and YANG (1962). The line of flight of an incident particle and the line of a produced particle establish a plane. If the produced particle decays into two particles, its decay products establish another plane. Since the pion has spin zero, it cannot carry information about the production plane, and therefore if a single pion is exchanged in the production reaction, there can be no correlation between the production and decay planes. As is usual with tests of particular models, although the presence of a correlation implies there must be corrections to the model, the absence of a correlation does not imply the model is correct.

[*] ISLAM and PIÑON found a strong $I = 0$ s-wave interaction (Ed.).

15. The ω meson

A. Mass and width

The ω meson was first observed by MAGLIĆ et al. (1961) as a three-pion correlation in antiproton-proton annihilation in the reaction

$$\bar{p} + p \to \pi^+ + \pi^- + \pi^0 + \pi^+ + \pi^- . \qquad (15.1)$$

See Fig. 15.1 for the three-pion peak superimposed on a phase space background. The mass and width of the ω are given by (GELFAND 1963)

$$M = 784.5 \pm 1\,\text{MeV}, \qquad \Gamma = 9.5 \pm 2.1\,\text{MeV} .$$

The principal decay mode is $\omega \to \pi^+ + \pi^- + \pi^0$. Assuming the width to be correct, the ω decays via strong interactions with conservation of isospin.

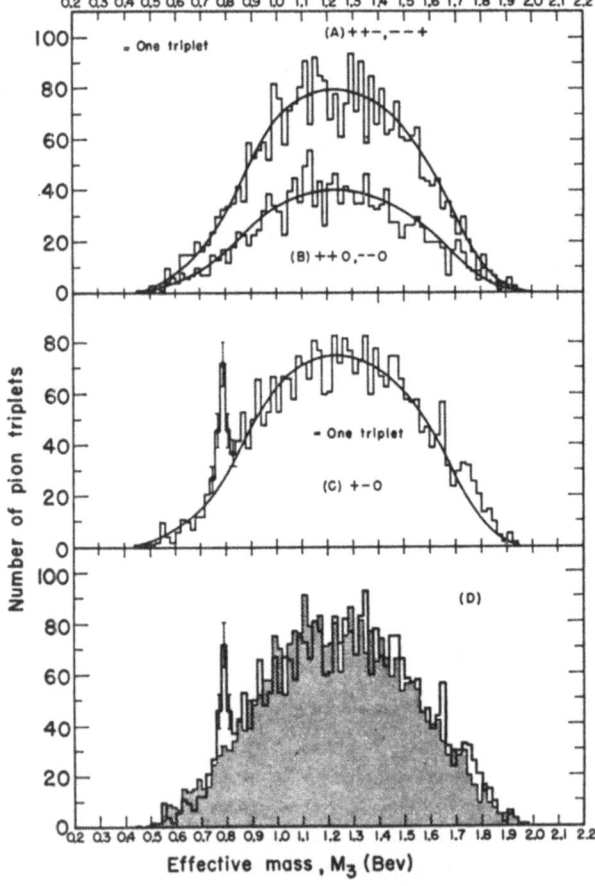

Fig. 15.1. Number of pion triplets versus effective mass (M_3) of the triplets for reaction $\bar{p} + p \to 2\pi^+ + 2\pi^- + \pi^0$. (A) is the distribution for the combination $|Q| = 1$; (B) is for the combination $|Q| = 2$ and (C) is for $Q = 0$ with 3200, 1600, and 3200 triplets respectively. In (D), the combined distributions (A) and (B) (shaded area) are contrasted with distribution (C) (heavy line) (MAGLIĆ 1961)

B. Isospin

Since only a neutral ω was observed, even though singly and doubly charged ω mesons were looked for, MAGLIĆ et al. concluded that the ω has isospin $I = 0$. This conclusion has been confirmed at other laboratories. For example ALFF et al. (1962) looked at ω mesons produced in the reaction

$$\pi^+ + p \rightarrow \omega + \Delta^{++}, \tag{15.2}$$

where the Δ (or $N^*_{3/2}$) is the pion-nucleon resonance with a mass of 1238 MeV and with quantum numbers $I = \frac{3}{2}$, $J^P = \frac{3}{2}^+$. The Δ will be discussed later (Section 21 B).

Since the initial state in reaction (15.2) is pure isospin $\frac{3}{2}$, the branching ratios into the different charge states are unique functions of the isospin I of the ω

$$
\begin{aligned}
\omega^+ : \omega^0 &= 0 & \text{if } I &= 0\,, \\
\omega^+ : \omega^0 &= 2:3 & \text{if } I &= 1\,, \\
\omega^{++} : \omega^+ : \omega^0 &= 2:2:1 & \text{if } I &= 2\,, \\
\omega^{+++} : \omega^{++} : \omega^+ : \omega^0 &= 20:10:4:1 & \text{if } I &= 3\,.
\end{aligned}
\tag{15.3}
$$

The mass spectrum of singly charged pion triplets did not show evidence of a peak, a fact that supports the assignment $I = 0$.

C. Spin and parity

To obtain information about the other quantum numbers of the ω, we need to look at the three pions from its decay. First, examine the consequences of the assumption that the ω has $I = 0$ and that isospin is conserved in the decay. Three pions in an $I = 0$ state have a unique isospin wave function, which is antisymmetric in the interchange of any two pions. This can be seen by noting that there is only one way to make a scalar (under rotation) out of three vectors: $\pi_1 \cdot \pi_2 \times \pi_3$. This wavefunction clearly changes sign under interchange of any two of the vectors.

Alternatively, one can look up the Clebsch-Gordan coefficients for combining three $I = 1$ particles to obtain $I = 0$. The wave function ψ_0 is

$$\psi_0 = (\pi_1^+ \pi_2^0 - \pi_1^0 \pi_2^+) \pi_3^- - (\pi_1^+ \pi_2^- - \pi_1^- \pi_2^+) \pi_3^0 + (\pi_1^0 \pi_2^- - \pi_1^- \pi_2^0) \pi_3^+ .$$

This can be seen to be antisymmetric under the interchange of any two pions. Inspection of this wave function also reveals that it has no all-neutral components. The experimental ratio of neutral to charged decays is

$$\Gamma(\omega \rightarrow \text{neutrals})/\Gamma(\omega \rightarrow \pi^+ \pi^- \pi^0) = 0.12 \pm 0.03\,, \tag{15.4}$$

a fact that suggests that I is conserved "most of the time".

A further check is that, since the isospin part of a three-pion wave function with $I = 0$ is antisymmetric, the space wave function must also be antisymmetric in the pion momenta to satisfy Bose statistics. This antisymmetric feature will also be present in the matrix element for the transition, and the square of the matrix element, as measured on a Dalitz plot, will be symmetric.

The transition matrix element will also depend on the spin and parity of the ω. Table 15.1 from STEVENSON et al. (1962) shows some simple examples of matrix elements with the proper symmetries. Note that

Table 15.1. Possible three-pion decays of a meson with $I^G = 0^-$, $J \leq 1$ (STEVENSON 1962). It is assumed that I and G are conserved in the decay. The matrix element is analyzed in terms of a pion pair with angular momentum L, and a third pion with angular momentum l in the 3-pion rest frame. Here E and p are the energy and momentum of a pion in the three-pion rest frame

Meson J^P	Matrix elements			
	l	L	Simple example	Vanishes at
1^-	1	1	$(p_0 \times p_+) + (p_+ \times p_-) \pm (p_- \times p_0)$	whole boundary
0^-	1,3	1,3	$(E_- - E_0)(E_0 - E_+)(E_+ - E_-)$	straight lines
1^+	0,2	1	$E_-(p_0 - p_+) + E_0(p_+ - p_-) + E_+(p_- - p_0)$	center, b, d, f

the matrix element has the opposite parity of the meson, since three pions have odd intrinsic parity. A Dalitz plot of the data of ALFF et al. (1962) is consistent with an antisymmetric matrix element, since the density of points vanishes where the relative momentum of any two pions is zero (see Fig. 15.2). This is evidence that the three pions are in

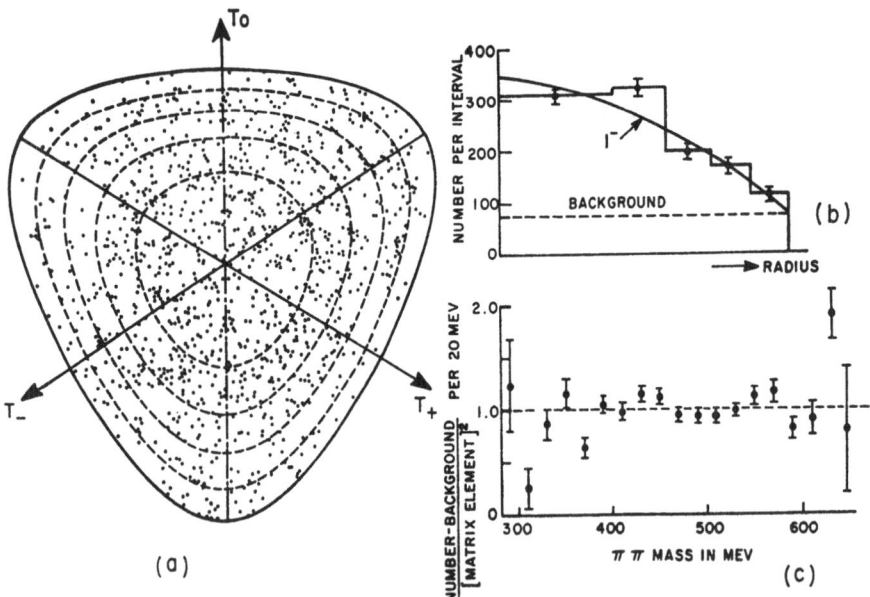

Fig. 15.2. (a) The Dalitz plot for 1100 omegas (including a background of 375 nonresonant triplets). (b) The density of points on the Dalitz plot compared to the expected density for a 1^- ω plus a uniformly distributed background. (c) The dependence of the $\pi\pi$ interaction in the $I = 1$ $J = 1$ state as a function of energy. This was obtained by summing the $\pi^+\pi^-$, $\pi^+\pi^0$, and $\pi^-\pi^0$ mass spectra for pion pairs from the ω decays, subtracting a background, and dividing by the distribution expected for 1^- decay into $\pi^+\pi^-\pi^0$. Since two of the three mass combinations are independent, an error corresponding to $(2/3N)^{1/2}$, where N is the number of pairs per interval before background subtraction, was assigned to each point (ALFF 1962)

an $I = 0$ state. Since a three-pion state has $G = -$, if $I = 0$, then $C = -$, as follows from the relation $C = (-1)^I G$. Furthermore, if the spin of the ω is less than two, the distribution of points on the Dalitz plot is consistent only with the assignment $J^P = 1^-$. This can be seen in Fig. 15.3 which shows the density of points as a function of radius on the Dalitz plot

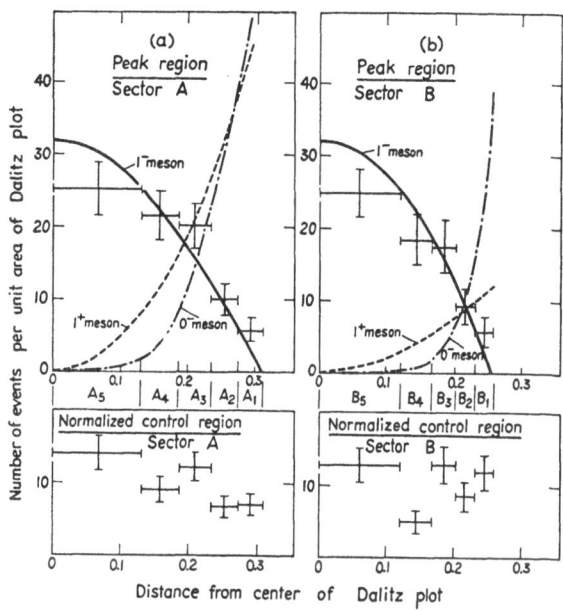

Fig. 15.3. Density of points on the Dalitz plot as a function of radius for 1^-, 0^-, and 0^+ mesons. Experimental points are also given. The Dalitz plot is divided into sectors A and B to test for the proper symmetry (STEVENSON 1962)

for 0^-, 1^- and 1^+. The agreement with the quantum numbers 1^- seems too striking to be coincidental. Thus the quantum numbers of the three-pion state, and therefore of the ω, are $I^G = 0^-$, $J^P = 1^-$, $C = -$. NAMBU (1957) predicted that a meson with these quantum numbers should exist in order to account for the isoscalar form factor of the nucleon.

D. Production of the ω

Because the ω has negative G parity, it cannot be produced in pion-nucleon collisions via the one-pion-exchange mechanism. However, it can be produced with the exchange of a ρ meson. Now the cross section for production of a particle via the mechanism of single meson exchange contains the factor $1/(q^2 + M^2)^2$, where M is the mass of the exchanged meson. Therefore the ω should be produced characteristically at larger momentum transfers than the ρ. This is the case, as seen from Fig. 14.2 (c) in which the production angular distribution of ω mesons is not peaked so much in the forward direction as that of the ρ.

E. Interference between ρ and ω

Experimental estimates of the branching ratio $R = \Gamma(\omega \to \pi^+\pi^-)/\Gamma(\omega \to \pi^+\pi^-\pi^0)$ have varied from $< 1\%$ to 15%. The decay $\omega \to \pi^+ + \pi^-$ is looked for as a sharp peak in the $\pi^+\pi^-$ mass spectrum at 785 MeV superimposed on the broad ρ peak*.

The difficulties which have plagued most experiments done so far are

1. The energy resolution has not been sufficiently good.
2. The sample of events has been too small for adequate statistics.
3. It is not completely clear how to interpret the results.

We discuss only the last of these three points. If an ω peak is superimposed on a ρ peak, the question arises as to whether there is any interference between the ρ and the two-pion decay of the ω. If interference is neglected, all events in the sharp peak are assumed to arise from two-pion decay of the ω, and the branching ratio R can be determined. However, if the *amplitudes* of the ω and ρ can interfere, a much smaller ω decay rate into 2 pions can lead to the same observed sharp peak. If the amplitudes are assumed to interfere, R depends on the width of the ω. Taking $\Gamma_\omega = 10$ MeV, and assuming completely coherent interference between the ρ and ω, WALKER et al. (1963) find

$$R = \Gamma(\omega \to \pi^+\pi^-)/\Gamma(\omega \to \pi^+\pi^-\pi^0) \approx 0.02 \ .$$

However, there is a difficulty with this explanation. If a ρ is produced primarily via a one-pion-exchange mechanism, it is produced with its spin perpendicular to the axis of the incident beam (in the rest system of the ρ). On the other hand, if an ω is produced primarily via ρ exchange, its spin is parallel to the axis of the incident beam (in the ω rest system). Therefore, there should not be any interference between the two-pion decays of the ρ and ω. ROSS (1964) has suggested that final state interactions might flip the spin of the ρ so that it can interfere with the ω.

F. Neutral decay of the ω

With the quantum numbers $J^{PC} = 1^{--}$, the ω should be able to decay into $\pi^0 + \gamma$. Presumably, this mode accounts for the neutral decays of the ω. However, if the rate $\Gamma(\omega \to \pi^0\gamma)/\Gamma(\omega \to \pi^+\pi^-\pi^0)$ is really 0.12 and if $\Gamma_\omega = 10$ MeV, then the partial decay rate of the ω into $\pi^0\gamma$ is

$$\Gamma(\omega \to \pi^0\gamma) \approx 1 \text{ MeV} \ .$$

This is the fastest electromagnetic decay known. One proposed explanation (BRONZAN 1963) is that the ω has odd A parity so that the decay $\omega \to \pi^0 + \gamma$ conserves A, while the electromagnetic decays of other mesons violate A-conservation. Note that the major decay mode $\omega \to \pi^+ + \pi^- + \pi^0$ also conserves A, while the decay $\omega \to \pi^+ + \pi^-$ violates both A and G conservation.

The decay $\omega \to \pi^+ + \pi^-$ is further inhibited relative to $\omega \to \pi^0 + \gamma$ because the rate for the former contains an extra power of $\alpha = 1/137$.

---

* If all data are combined, there is no statistically significant indication for the ω^0 decay into 2 pions (LÜTJENS 1964) (Ed.).

However, the two-pion decay rate is enhanced because of the ρ resonance. The net result is therefore difficult to estimate. Detailed theoretical calculations of these rates have been made, but will not be discussed here.

If the ω has odd A parity and the photon has even A parity, as BRONZAN and LOW suggest, then the coupling of the ω to the photon is small, and the ω is not responsible for the isoscalar form factor of the nucleon. There is another candidate to serve this function, however, the φ meson (to be discussed later).

16. The η meson

A. Mass and width

The η meson (sometimes called the χ) was discovered by PEVSNER et al. (1961) by looking at the interactions of pions of momentum 1.23 GeV/c in a deuterium bubble chamber.

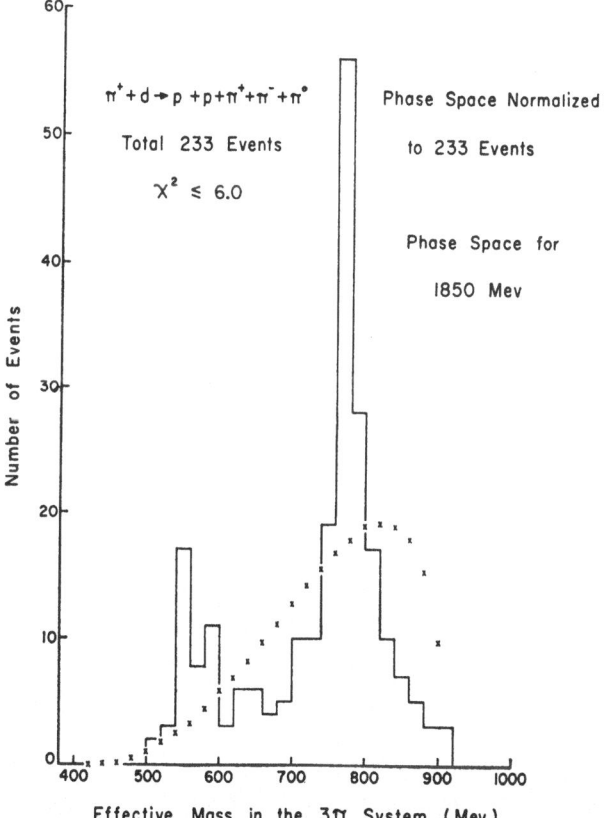

Fig. 16.1. Histogram of the effective mass of the three-pion system for 233 events, showing evidence for the η and the ω mesons. The crosses show the phase space distribution (PEVSNER 1961)

The η is a neutral particle, so its existence can be inferred either from conservation of energy and momentum or from observation of its decay into other particles. In fact, the η was observed from its decay into three pions: $\eta \to \pi^+ + \pi^- + \pi^0$. Since the η did not travel a measurable distance in the bubble chamber before decaying, the existence of the η was deduced from a three-pion energy correlation in the reaction

$$\pi^+ + d \to p + p + \pi^+ + \pi^- + \pi^0 . \tag{16.1}$$

The ω meson was also seen in this experiment as a prominent peak in the invariant mass distribution of the three pions. See Fig. 16.1 (PEVSNER 1961).

The η has a mass

$$M = 548.5 \pm 0.6 \text{ MeV}$$

and a measured width

$$\Gamma < 7 \text{ MeV} .$$

B. Isospin

Evidence that the η has $I = 0$ is the absence of a peak in the three-pion mass spectrum in the reaction

$$\pi^+ + d \to p + n + \pi^+ + \pi^+ + \pi^- .$$

This fact alone is not sufficient to rule out $I = 1$ for two reasons:

1. Although the initial state $(\pi^+ d)$ is a pure $I = 1$ state, the final state $(p n \pi^+ \pi^+ \pi^-)$, considered as a state of p, n and a hypothetical η^+, does not have a unique isospin configuration. Therefore, interference effects could inhibit the production of η^+ if it existed.

2. The η^+ might exist but not decay appreciably into the mode $\pi^+ \pi^+ \pi^-$.

However, by now both these objections have been overcome. Evidence against the first possibility is that the η has been seen in numerous experiments, including the quasi-two-particle mode (ALFF 1962)

$$\pi^+ + p \to \Delta^{++} + \eta^0 ,$$

where Δ is the pion-nucleon resonance at a mass of 1238 MeV. In this reaction, the branching ratio into the various charge states is a unique function of the isospin, and no evidence for a $\pi^+ \pi^+ \pi^-$ correlation has been seen.

Experiment has also shown that the second objection is invalid. For example, CARMONY, ROSENFELD and VAN DE WALLE (1962) have looked without success for a peak at 550 MeV in the spectrum of $\pi^\pm$ + neutrals produced in the reaction

$$\pi^\pm + p \to p + \pi^\pm + \text{neutrals}$$

at an incident momentum of 1.25 GeV/c. This experiment was done at essentially the same momentum as the experiment of PEVSNER et al. Therefore, by charge independence, if the $\eta^\pm$ existed it would have been seen.

C. Spin, parity and G parity

Good evidence that the η has $J^{PG} = 0^{-+}$ comes from a Dalitz plot. As with the ω, the matrix element for the decay depends on the quantum numbers. We have already listed some matrix elements in Table 15.1 in connection with the ω decay. In Table 16.1 are listed some other matrix

Table 16.1. The G-forbidden three-pion decays of an $I = 0$ meson (BASTIEN 1962a). The matrix element is analyzed in terms of a $\pi^+\pi^-$ pair with relative momentum $\boldsymbol{p}$ and angular momentum L and a π^0 with momentum $\boldsymbol{q}$ and angular momentum l in the 3-pion rest frame. Here T_0 is the π^0 kinetic energy

Meson J^{PG}	l	L	Simplest matrix element	Vanishes at	Dominant radiative decay modes
0^{-+}	0	0	1	nowhere	$2\gamma, \pi^+\pi^-\gamma$
1^{++}	1	0	1	$T_0 = 0$	$\pi^+\pi^-\gamma$
1^{-+}	2	2	$(\boldsymbol{p} \times \boldsymbol{q})(\boldsymbol{p} \cdot \boldsymbol{q})$	T_0 axis and boundary	$\pi^+\pi^-\gamma$

elements which are not antisymmetric in the momenta of the three pions because the pions are not in an $I = 0$ state (BASTIEN 1962a).

Fig. 16.2 (ALFF 1962) shows a Dalitz plot for η decays. Here the π^0 kinetic energy T_0 is plotted against the difference in kinetic energy $(T_+ - T_-)/\sqrt{3}$ of the π^+ and π^-. Only $1/2$ of the kinematically allowed region is shown in Fig. 16.2 for the following reason:

The η presumably has a definite C parity, and C is conserved in its decay. Therefore, the final state consisting of $\pi^+\pi^-\pi^0$ must also have a

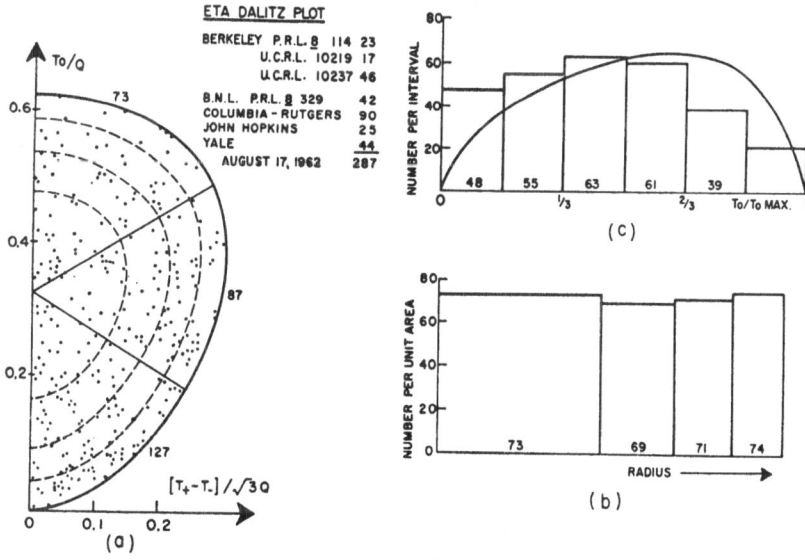

Fig. 16.2. The Dalitz plot and projections for published η decays into $\pi^+\pi^-\pi^0$. (a) shows the distribution of points, (b) the radial density, and (c) the projection of the points on the π^0 axis. The solid line in (c) corresponds to uniform population (ALFF 1962)

definite C parity. This implies that the square of the final state 3-pion wave function must be symmetric under the interchange of the π^+ and π^-. It is actually observed that both halves of the Dalitz plot are indistinguishable within experimental error. Therefore $(T_+ - T_-)/\sqrt{3}$ is actually plotted in Fig. 16.1 to obtain double the number of events.

From Fig. 16.2 it is seen that the density of points does not vanish along the boundary, along the T_0 axis, or at $T_0 = 0$ as it should for all the assignments considered in Tables 15.1 and 16.1 except 0^{-+}. This fact strongly supports the assignment $J^{PG} = 0^{-+}$. It is seen from Table 16.1 that the simplest matrix element for $J^{PG} = 0^{-+}$ has uniform density. The observed density is clearly not uniform, as there is a depopulation of events at large T_0. The depopulation of this region is in no way inconsistent with the assignment $J = 0^{-+}$, but the deviation from uniformity means that there are final state interactions among the emitted pions.

Other evidence for the assignment $J^{PG} = 0^{-+}$ comes from the observation by CHRÉTIEN et al. (1962) of the decay $\eta \to 2\gamma$. This shows that the η has $C = +$, and therefore $G = +$, since for an $I = 0$ meson, $G = C$. This observation also rules out $J = 1$, since a spin 1 particle cannot decay into two photons.

With the quantum numbers 0^{-+}, the η has the same quantum numbers as a four-pion state. It does not have enough mass to decay into four charged pions, and decay into four neutral pions is severely inhibited by lack of available phase space.

It is interesting that, with the assignment 0^{-+}, the η cannot be produced in peripheral collisions with the exchange of any single definitely established meson. Experiment confirms that the η is not produced preferentially in low momentum transfer events. This is illustrated in Fig. 14.2 (b) (ALFF 1962) which shows a relatively isotropic production angular distribution. Also in Fig. 14.2 are shown the production angular distributions for ω and ρ. The ρ, which can be produced in an event with single pion exchange, is peaked strongly in the forward direction. Production of the ω, which can occur via single ρ exchange, requires a somewhat larger momentum transfer, and the forward peaking is not so pronounced.

D. Decay modes

The relative decay rates of a number of different decay modes of the η have been directly measured. The measured branching ratios are (CRAWFORD 1963)

$$\Gamma(\eta \to \pi^+ \pi^- \gamma)/\Gamma(\eta \to \pi^+ \pi^- \pi^0) = 0.26 \pm 0.08 , \tag{16.2}$$

$$\Gamma(\eta \to 3\pi^0)/\Gamma(\eta \to \pi^+ \pi^- \pi^0) = 0.83 \pm 0.32 , \tag{16.3}$$

$$\Gamma(\eta \to \pi^+ \pi^- \gamma)/\Gamma(\eta \to 2\gamma) = 0.21 \pm 0.12 , \tag{16.4}$$

$$\Gamma(\eta \to 2\gamma)/\Gamma(\eta \to \pi^+ \pi^- \pi^0) = 1.25 \pm 0.56 . \tag{16.5}$$

From these numbers we obtain

$$\Gamma(\eta \to \text{neutrals})/\Gamma(\eta \to \text{charged}) = 1.65 \pm 0.53 . \tag{16.6}$$

This number should be compared to the number obtained by counting missing neutrals (BASTIEN 1962a, FOELSCHE 1962, ALFF 1962)

$$\Gamma(\eta \to \text{neutrals})/\Gamma(\eta \to \text{charged}) = 2.7 \pm 0.6 . \tag{16.7}$$

The agreement between (16.6) and (16.7) is only fair.

The branching ratios of the η into its various modes present a puzzle. All the observed decays presumably go via the electromagnetic interaction. The decay rates $\eta \to \pi^+ + \pi^- + \pi^0$ and $\eta \to 2\gamma$ are both proportional to α^2 where α is the fine structure constant ($\alpha = 1/137$). But the phase space for the 2γ decay is two orders of magnitude greater than the phase space for 3π decay. Therefore the decay $\eta \to 2\gamma$ should dominate, but the observed rates are comparable (Eq. 16.5). Two different kinds of explanations have been given for this fact [*]:

1. The decay $\eta \to 2\gamma$ is forbidden by conservation of A parity (OKUBO 1963, BRONZAN 1963) just as the decay $\pi^0 \to 2\gamma$ is forbidden.

2. The decay $\eta \to 3\pi$ is enhanced by final state pion-pion interactions, perhaps by a resonance in the pion-pion system (BROWN 1962, 1964).

The lifetime of the η has not been measured. We merely quote one attempt to estimate it, that of BROWN and SINGER (1962)

$$\Gamma(\text{theory}) \approx 250 \text{ eV}, \quad \tau(\text{theory}) \approx 3 \times 10^{-18} \text{ sec} .$$

This is two orders of magnitude shorter than the measured π^0 lifetime.

17. The K* meson

A. Mass and width

The K* was first seen by ALSTON et al. (1961) in the reaction

$$K^- + p \to \overline{K}{}^0 + \pi^- + p \tag{17.1}$$

at an incident K^- momentum of 1.15 Gev/c. Its mass and width are

$$M = 888 \pm 3 \text{ MeV}, \qquad \Gamma = 50 \pm 10 \text{ MeV} .$$

A Dalitz plot of the $K\pi N$ system is shown in Fig. 17.1 (ALSTON 1961). It is apparent from Fig. 17.1 that the proton likes to come off at an almost unique energy of ≈ 20 MeV. This energy corresponds to a mass $M_{K*} = 890$ MeV. Also shown in Fig. 17.1 is a line on the Dalitz plot corresponding to formation of the Δ (or $N_{3/2}^*$). Since there is no clustering of phase points around this line, there is no interference between K* and Δ production in this reaction.

---

[*] Further references: BARRET et al. (1964), CHAM (1964) (Ed.).

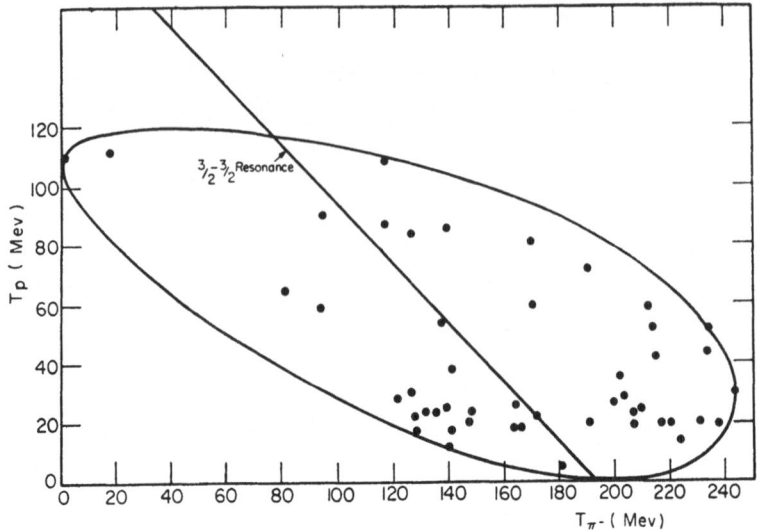

Fig. 17.1. Phase-space plot of 48 examples of $K^- + p \rightarrow \overline{K}^0 + \pi + p$ reactions, showing evidence for the K* meson (ALSTON 1961)

The quantum numbers of the K* have been determined

$$Y = 1, \qquad I = {}^1/_2, \qquad J^P = 1^- .$$

B. Isospin

The branching ratio

$$R = \Gamma(K^{*-} \rightarrow \overline{K}^0 + \pi^-)/\Gamma(K^{*-} \rightarrow K^- + \pi^0) \qquad (17.2)$$

is predicted to be

$$R = 2 \quad \text{if} \quad I = {}^1/_2; \quad R = {}^1/_2 \quad \text{if} \quad I = {}^3/_2 . \qquad (17.3)$$

The experimental ratio agrees with $I = {}^1/_2$. This conclusion depends on I being conserved in the decay, a reasonable assumption since the width of the K* shows that the decay is allowed by strong interactions. In support of the assignment $I = {}^1/_2$ is the fact that doubly charged K* mesons have not been seen.

C. Spin and parity

CHINOWSKY et al. (1962) have seen the K* produced in the reaction

$$K^+ + p \rightarrow p + K^+ + \pi^- + \pi^+ . \qquad (17.4)$$

Nearly all the events were consistent with the reaction $K^+ + p \rightarrow K^{*0} + \Delta^{++}$. The production angular distribution of the K* is peaked forward and fits the prediction of the peripheral model with a one-pion-exchange diagram. Because of the forward peaking of the K*, an Adair analysis is useful. Examining events with production angle $\cos\Theta > 0.8$, CHINOWSKY et al. (1962) found the decay angular distribution to be

$$I_{K*}(\theta) \approx \cos^2\theta . \qquad (17.5)$$

This shows immediately that $J > 0$.

Since the Δ has spin $J_\Delta = {}^3/_2$, the Adair analysis does not give a unique result. Assume, however, that $J_{K*} = 1$. Then the possibilities for the angular distributions of the decays of the K* and Δ are given in Table 17.1 (CHINOWSKY 1962). Intermediate cases in which the K*

Table 17.1. Some possible decay angular distributions of the K* and Δ produced in the reaction $K^+ + p \to K^* + \Delta$, assuming the K* has spin one (CHINOWSKY 1962). The quantum numbers m_p, m_{K*}, and m_Δ are the projections of the spins of the proton, K*, and Δ along the beam direction, and the decay angular distributions $I_{K*}(\theta)$ and $I_\Delta(\theta)$ are measured with respect to the beam axis. The protons in the target are unpolarized, but the angular distributions are unchanged if all the m values are replaced by their negatives

m_p	m_{K*}	m_Δ	$I_{K*}(\theta)$	$I_\Delta(\theta)$
${}^1/_2$	0	${}^1/_2$	$\cos^2\theta$	$1 + 3\cos^2\theta$
${}^1/_2$	1	$-{}^1/_2$	$\sin^2\theta$	$1 + 3\cos^2\theta$
${}^1/_2$	-1	${}^3/_2$	$\sin^2\theta$	$\sin^2\theta$

and Δ are not in pure m-states are of course also possible. The observed angular distributions are

$$I_{K*}(\theta) = \cos^2\theta, \qquad I_\Delta(\theta) = 1 + 3\cos^2\theta \qquad (17.6)$$

in agreement with one of the possibilities in Table 17.1, a fact which supports the assignment $J_{K*} = 1$. These same angular distributions are predicted by the one-pion-exchange model.

2. There is other evidence not depending on the one-pion-exchange model, which supports $J = 1$. ALSTON et al. (1961) obtained a decay angular distribution suggesting $J \leq 1$. The argument goes as follows: The $\overline{K}*$ was produced in the reaction $K^- + p \to \overline{K}* + p$ at an incident K^- momentum of 1.15 GeV/c. Since this corresponds to a low c.m. momentum of the $\overline{K}*$ and since the production angular distribution was isotropic, the assumption was made that the production was predominantly in the s state. If this assumption is correct, the Adair analysis can be applied rigorously to all events. The $\overline{K}*$ decay angular distribution (for $J > 0$) is then given by

$$I(\theta) = |a\,Y_J^0|^2 + |b\,Y_J^1|^2 , \qquad (17.7)$$

where $|a|^2 + |b|^2 = 1$.

The second term in this formula arises from the fact that the proton spin can be flipped. The angular distribution for $J = 1$ can vary anywhere between $\sin^2\theta$ and $\cos^2\theta$, depending on the relative magnitudes of the proton spin-flip and non-spin-flip amplitudes. However, for $J > 1$, the K* must be partially aligned (m-values > 1 are forbidden) and the decay cannot be isotropic. These arguments are quantitatively summarized in the following expression for the average value of $\cos^2\theta$ as a function of the spin of the K*

$$\langle\cos^2\theta\rangle = \frac{2J^2 + 2J - 3 + 2|a|^2}{4J^2 + 4J - 3}, \tag{17.8}$$

where

$$\langle\cos^2\theta\rangle = \int I(\theta)\cos^2\theta\,d(\cos\theta)/\int I(\theta)\,d(\cos\theta).$$

In particular, if $J = 2$, then $\langle\cos^2\theta\rangle \geq 0.429$; and the minimum value of $\langle\cos^2\theta\rangle$ increases as J increases. The experimental result

$$\langle\cos^2\theta\rangle = 0.275 \pm 0.051 \tag{17.9}$$

is three standard deviations from being consistent with $J = 2$, but is consistent with $J = 0$ or 1. Combined with the result of CHINOWSKY et al., $J > 0$, this is evidence for $J = 1$.

3. Assuming parity is conserved in the strong decay $K^* \rightarrow K + \pi$, we immediately have the result

$$P_{K^*} = -P_K(-1)^J \tag{17.10}$$

since the intrinsic parity of the π is negative and the parity of orbital angular momentum is $(-1)^J$. Because of this relation, the relative parity of the K* and K is determined by a measurement of the K* spin. Since $J = 1$, we have $P_{K^*} = P_K$. Since the K parity is measured to be negative, so is the parity of the K*.

4. CALDWELL (1961) has suggested as an application of the Bohr analysis a method to obtain information about the spin and parity of the K*. Actually, there are practical difficulties with this method, and, as we have seen, the spin and parity of the K* have already been measured. Nevertheless, we shall give this application as an illustration of the Bohr analysis. Consider the reaction

$$K + He^4 \rightarrow K^* + He^4. \tag{17.11}$$

Since the K and He^4 both have spin 0, Bohr's relation becomes

$$P_K = P_{K^*}(-1)^{J_n} \tag{17.12}$$

where J_n is the component of the K* spin which is normal to the production plane. Comparing this relation with the expression for the K* parity previously obtained

$$P_{K^*} = -P_K(-1)^J \tag{17.13}$$

we see that $J = 0$ implies both $P_K = P_{K^*}$ and $P_K = -P_{K^*}$. Thus, if parity is conserved, the reaction is forbidden.

Assuming parity conservation, if the reaction is seen, then $J \geq 1$. For $J = 1$ we must have $J_n = 0$ which yields a pure $\cos^2\theta_n$ decay angular distribution, where θ_n is the angle the π makes with the normal in the rest system of the K*. For $J = 2$ we must have $J_n = 1$, and the decay angular distribution is $\cos^2\theta_n \sin^2\theta_n$. For $J \geq 3$, the angular distribution cannot be uniquely predicted.

The practical disadvantage is that the reaction $K + He^4 \rightarrow K^* + He^4$ is likely to be rare regardless of the value of the K* spin (CALDWELL 1961). This can be seen as follows: Consider the cases in which the K* is produced in the beam direction. Then, by the Adair argument, the decay angular distribution of the K* is $|Y_J^0|^2$. But $|Y_J^0|^2$ has a finite amplitude along the beam direction, so that if the K* is produced at $\Theta = 0$, there will be a finite amplitude for $K + He^4 \rightarrow He^4 + K + \pi$ with all particles produced along the beam direction. But this configuration is forbidden by conservation of parity, as can be seen by performing a reflection with respect to the beam axis. Since the He^4, K and π have spin 0 and there is no orbital angular momentum about the beam direction, the only change as a result of the reflection is a change in parity because the pion is pseudoscalar. This means that the K* must be produced only in state of Y_L^m with $m \neq 0$, i.e., principally at angles $\Theta > 1/L_{max}$. But at these angles the momentum transfer to He^4 is sufficient to break it up in most cases, so the cross section is likely to be small. Because of this difficulty, CALDWELL also discussed

$$K + He^4 \rightarrow K^* + \pi + He^4 . \tag{17.14}$$

See his paper for details.

5. Still another method to measure the spin of the K* has been proposed by SCHWARTZ (1961). Again we discuss this method partly because it is an interesting application of the conservation laws, and partly because it provides additional evidence that $J \neq 0$.

Consider antiproton-proton annihilation at rest into the modes

$$\bar{p} + p \rightarrow \bar{K}^0 + K^{*0} \quad \text{and} \quad \bar{p} + p \rightarrow K^0 + \bar{K}^{*0} \tag{17.15}$$

with subsequent decays

$$K^* \rightarrow K^0 + \pi^0 \quad \text{and} \quad \bar{K}^* \rightarrow \bar{K}^0 + \pi^0 . \tag{17.16}$$

This method consists of determining whether the K^0 and $\bar{K}^0$ (one of which comes from the decay of a K* or a $\bar{K}^*$) decay into $K_1 K_1$, $K_2 K_2$ or $K_1 K_2$. It is necessary to assume that the $\bar{p}p$ annihilation occurs in an s state. We have already discussed the basis for this assumption.

Now an s state of $\bar{p}p$ has negative parity and total angular momentum zero or one. If L is the orbital angular momentum of the K with respect to the K* and J is the K* spin, the parity of the $K\bar{K}^*$ state is

$$P = P_{K^*} P_K (-1)^L = (-1)^{L+J+1} = - . \tag{17.17}$$

This implies $L + J$ is even, which can occur only if $L = J$, because the total angular momentum is zero or one.

First consider the case in which the spin of the K^* is $J = 0$. Then $L = 0$, and $\bar{p}p$ annihilation must occur from the singlet state. But in this state $CP(\bar{p}p) = -$. Now consider the states $K_1 K_1 \pi^0$, $K_1 K_2 \pi^0$, and $K_2 K_2 \pi^0$ with $L = J = 0$. Of these states, $K_1 K_1 \pi^0$ and $K_2 K_2 \pi^0$ are odd under CP and $K_1 K_2 \pi^0$ is even, so that $J = 0$ implies that only the decays $K_1 K_1 \pi^0$ or $K_2 K_2 \pi^0$ are allowed. If $J \neq 0$, in general all three modes will occur. Experimentally, $K_1 K_2$ pairs have been seen with one of the pair having a unique energy corresponding to a $\bar{K}^0 K^{*0}$ state. This is evidence for $J \neq 0$.

It is convenient to discuss here how the modes $K_1 K_1$, $K_1 K_2$ and $K_2 K_2$ are distinguished. The K_2 has a lifetime $\approx 6 \times 10^{-8}$ sec, which is sufficiently long so that the K_2 leaves the detection area before decaying. Thus, K_2 decays are not seen, and the mode $K_2 K_2$ is not detected. The K_1 lifetime, on the other hand, is sufficiently short (10^{-10} sec) so that if it decays into $\pi^+ \pi^-$ (which it does $2/3$ of the time) it is usually seen*. In the $\bar{p}p \to \bar{K} K^*$ reaction, one concentrates on events in which a visible K_1 has a unique energy. Then, if the K^* decays into $K_1 + \pi$, this K_1 should be visible $2/3$ of the time (with 100% detection efficiency). The experimentally observed branching ratio is (ARMENTEROS 1962)

$$\frac{K_1(\text{seen})\, K_1(\text{seen})}{K_1(\text{seen})\, K(\text{not seen})} = \frac{13 \pm 11}{43 \pm 14} \tag{17.18}$$

compared to the value 2 which would be predicted for $K_1 K_1$. This strongly indicates that some $K_1 K_2$ are being produced and that therefore the K^* has $J \neq 0$.

18. The φ meson

A. Mass and width

The φ meson was observed by BERTANZA et al. (1962a) in the reactions

$$K^- + p \to \Lambda + K^0 + \bar{K}^0 ,$$
$$\to \Lambda + K^+ + K^- .$$

with 2.24 and 2.5 GeV/c incident K^- mesons. A Dalitz plot, showing the peak in the $K\bar{K}$ mass distribution is shown in Fig. 18.1 (BERTANZA 1962a). Subsequent experiments by SCHLEIN et al. (1963) and by CONNOLLY et al. (1963a) have established the quantum numbers of the φ to be $J^{PC} = 1^{--}$. Its isospin and G parity are $I^G = 0^-$. SAKURAI (1962) predicted that these should be the quantum numbers of the φ on the basis of the observation of the decay mode $\varphi \to K_1 + K_2$.

---

* How often the decay is seen depends on the efficiency of detection, which is presumably known.

The mass and width of the φ are (GELFAND 1963a, CONNOLLY 1963a, SCHLEIN 1963)

$$M = 1019.5 \pm 0.5 \text{ MeV}, \qquad \Gamma = 3.1 \pm 0.8 \text{ MeV}.$$

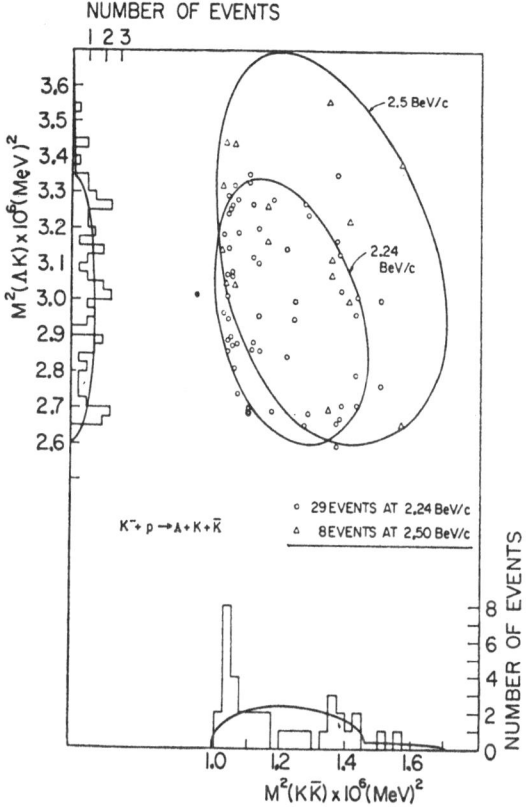

Fig. 18.1. The Dalitz plot of $\Lambda K \bar{K}$ events, showing evidence for the φ meson. The solid curves on the projections are the invariant phase space curves normalized to the total number of events (BERTANZA 1962a)

B. Spin, parity and C parity

Evidence concerning the spin and parity of the φ comes from observation of its decay modes

$$\begin{aligned} \varphi &\to K^+ + K^- \\ &\to K_1 + K_2 \end{aligned} \tag{18.1}$$

with a branching ratio (CONNOLLY 1963a)

$$R = \frac{\Gamma(\varphi \to K_1 K_2)}{\Gamma(\varphi \to K_1 K_2) + \Gamma(\varphi \to K^+ K^-)} = 0.45 \pm 0.10 \tag{18.2}$$

1. Observation of the mode $\varphi \to K_1 K_2$ shows that the φ has odd spin, negative parity, and negative C parity. These assignments follow

from the relations

$$CP(K^0\overline{K}^0) = +, \qquad CP(K_1K_2) = (-1)^{J+1}$$
$$C(K^+K^-) = P(K^+K^-) = (-1)^J . \tag{18.3}$$

The assignments $C = P = -$ for the φ depend on the assumption that C and P are conserved in its decay. In view of the 3 MeV width of the φ, this is a reasonable assumption. The conclusion that J is odd depends only on CP conservation. With the quantum numbers J odd, $P = C = -$, the neutral decay is allowed to proceed into the mode $\varphi \to K^0 + \overline{K}^0$ via strong interactions with the K^0 and $\overline{K}^0$ subsequently decaying into K_1K_2.

2. Independently of whether the φ has isospin zero or one, the branching ratio R (Eq. 18.2) is given by $R = 0.5$ if isospin is strictly conserved in the decay. However, R depends on the spin because of the mass difference between the K^0 and K^+ and because of Coulomb effects in the K^+K^- decay. These effects depend on the assumed interaction radius for the decay process. Choosing an interaction radius of $1/2M_\pi$, we obtain $R(J = 1) = 0.39$ and $R(J = 3) = 0.26$. The experimental value of R favors $J = 1$ by 2 standard deviations. SCHLEIN et al. (1963) obtained essentially the same results with somewhat fewer events.

3. The decay angular distribution of the φ has been observed to be anisotropic. The Adair analysis does not lead to a unique prediction. This is because in the production reaction the z component of the Λ spin may be either the same or opposite (spin-flip amplitude) to the initial proton spin. The decay angular distribution for spin J is given by

$$I(\theta) = |Y_J^0(\theta)|^2 + \varrho^2 |Y_J^1(\theta)|^2 \tag{18.4}$$

where ϱ is the ratio of the spin-flip to the non spin-flip production amplitude. Taking into account parity non-conservation in the decay of the Λ, one may obtain additional information from angular correlations between the decay of the Λ and decay of the φ (TREIMAN 1962a). In an experiment of SCHLEIN et al. (1963), the major information comes from the Adair analysis at production angles with $\cos\Theta \geq 0.5$. The data favor $J = 1$ over $J = 3$ by four to one odds, and are inconsistent with $J = 0$ by two standard deviations.

Additional evidence (CONNOLLY 1963a) for $J = 1$ comes from looking at the decay angular distribution of the φ with respect to the angle ξ. It is given by
$$I(\xi) = 1 - (0.64 \pm 0.30) \cos^2\xi .$$

Here ξ is the azimuthal angle in a coordinate system in which the z axis is the incident direction and the x axis is the normal to the production plane in the φ rest system.

C. Isospin and G parity

Evidence that the φ has isospin $I = 0$ comes from observation of the reactions (CONNOLLY 1963a)

$$K^- + p \to \Sigma^0 + K^+ + K^- \to \Sigma^+ + K^- + K^0 \tag{18.5}$$
$$\to \Sigma^- + K^+ + \overline{K}^0 .$$

A sharp peak in the K^+K^- mass spectrum was observed at the mass of the φ, but no corresponding peaks were observed in the $K^+\bar{K}^0$ or K^-K^0 spectra.

An unanswered question is whether the K^+K^- enhancement arises from the decay of the φ or whether it is some other effect in the $I = 0$ state. There is, in fact, some evidence that there is an $I = 0$ $K\bar{K}$ enhancement in a state with J even (see Section 20D). This latter effect leads to an enhancement of the K_1K_1 mass spectrum as well. CONNOLLY et al. (1963a) did not report whether they observed any K_1K_1 events. (The mass spectrum of the mode K_1K_2 is not obtainable in the reaction $K^-p \rightarrow \Sigma^0 K_1 K_2$, since two unobserved particles occur in the final state.)

However, let us assume that all the K^+K^- pairs come from φ decay. Consider the ratio R, defined by

$$R = (\Sigma^- K^+ \bar{K}^0 + \Sigma^+ K^- K^0)/(\Sigma^0 K^+ K^-) . \tag{18.6}$$

Clearly, if the φ has $I = 0$, then $R = 0$. Now suppose the φ has $I = 1$. Then, since the initial K^-p state is a mixture of $I = 0$ and $I = 1$, the branching ratios into the various channels cannot be predicted. However, R satisfies the inequality $R \geq 4$ (assuming that half the φ^0's decay into K^+K^-). The observed value of R for events above a very small background is $R=0$ in agreement with $I=0$ and in disagreement with $I=1$.

If $I = 0$, then $G = -$, since $G = (-1)^I C$ and C was shown to be odd. In agreement with this assignment is the absence of the decay mode $\varphi \rightarrow \pi^+ + \pi^-$. The observed branching ratio is (CONNOLLY 1963a)

$$\Gamma(\varphi \rightarrow \pi^+\pi^-)/\Gamma(\varphi \rightarrow K\bar{K}) < 0.08 \tag{18.7}$$

although the estimated rate would be about 10 if $G = +$ and the couplings were the same for both decays. However, if G is negative, then the decays $\varphi \rightarrow 3\pi$ and $\varphi \rightarrow \rho\pi$ should occur. But the decay into $\rho + \pi$ is also inhibited. The observed branching fraction is

$$\Gamma(\varphi \rightarrow \rho\pi)/\Gamma(\varphi \rightarrow K\bar{K}) = 0.10 \pm 0.10 \tag{18.8}$$

although an estimate based on phase space is that $\varphi \rightarrow \rho + \pi$ should be the dominant decay mode.

According to the experimental evidence, the φ has the same quantum numbers as the ω: $I^G \doteq 0^-$, $J^P = 1^-$. But the decay width of the φ is only $1/3$ the decay width of the ω: $\Gamma(\varphi) = 3.1$ MeV, $\Gamma(\omega) = 9.5$ MeV. The difference in these widths cannot be explained on the basis of phase space alone, since the phase space for the ω to go into 3 pions is much smaller than the corresponding phase space for the φ. Also the φ has available the modes $\varphi \rightarrow K\bar{K}$ and $\varphi \rightarrow \rho + \pi$ which are forbidden to the ω by energy conservation. Yet the ω has a larger width.

BRONZAN and LOW have interpreted this in terms of conservation of A parity. If the ω is assigned negative A parity and the φ positive A parity, the decays $\varphi \rightarrow \rho + \pi$ and $\varphi \rightarrow 3\pi$ are forbidden. If A parity is a good quantum number, it is the φ, rather than the ω, which is chiefly responsible for the nucleon isoscalar form factor.

Another interpretation is in terms of $\omega - \varphi$ "mixing". Since these particles have the same I^G, J^P, they may interact in such a way that one of them has a much stronger interaction with 3 pions than the other. We shall not go into the details of this question, but merely refer to one of many theoretical treatments, that by KATZ and LIPKIN (1963).

19. Some other mesons

A. The f meson

Clear evidence for the existence of the f meson was first obtained by SELOVE et al. (1962) although preliminary evidence was seen earlier. The f appeared as a peak in the $\pi^+\pi^-$ mass spectrum in the reaction

$$\pi^- + p \rightarrow n + \pi^+ + \pi^- \tag{19.1}$$

with 3.0 GeV/c incident pions. Since the peak was not observed in the reaction

$$\pi^- + p \rightarrow p + \pi^- + \pi^0, \tag{19.2}$$

the authors concluded that the f has isospin $I = 0$. Histograms showing the $\pi^-\pi^0$ and $\pi^+\pi^-$ mass spectra are shown in Fig. 19.1. In the $\pi^-\pi^+$ spectrum, two peaks are observed, one at the mass of the ρ^0, and the other, the f meson, at a mass

$$M_f = 1250 \text{ MeV}, \qquad \Gamma = 120 \text{ MeV} .$$

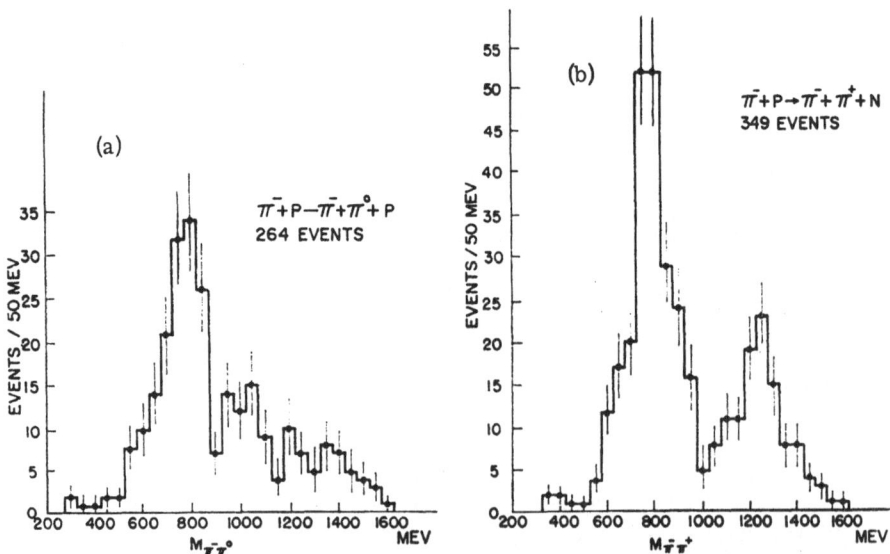

Fig. 19.1. (a) The $\pi\pi$ mass plot for the reaction $\pi^- + p \rightarrow \pi^- + \pi^0 + p$. (b) The $\pi\pi$ mass plot for the reaction $\pi^- + p \rightarrow \pi^- + \pi^+ + n$, showing evidence for the f meson (SELOVE 1962)

It can also be seen from Fig. 19.1 that only the ρ^- peak is observed in the $\pi^-\pi^0$ spectrum. The fact that no charged f is seen is not proof that $I = 0$, as the initial state is a coherent mixture of $I = {}^3/_2$ and $I = {}^1/_2$. Therefore, interference effects could suppress the production of an f^- if it existed. GUIRAGOSSIÁN (1963) has also observed the f^0, at a slightly higher incident pion momentum (3.3 GeV/c) and he also did not see a charged f. Assuming that the f has isospin zero, then its spin must be even and its parity positive because the pions are bosons.

The c.m. decay angular distributions of the $\pi^-\pi^0$ and $\pi^-\pi^+$ are shown in Fig. 19.2. The ρ^- decays symmetrically about $\theta = 90°$ [Fig. 19.2(a)], but the ρ^0 does not, with the π^- tending to go forward [Fig.

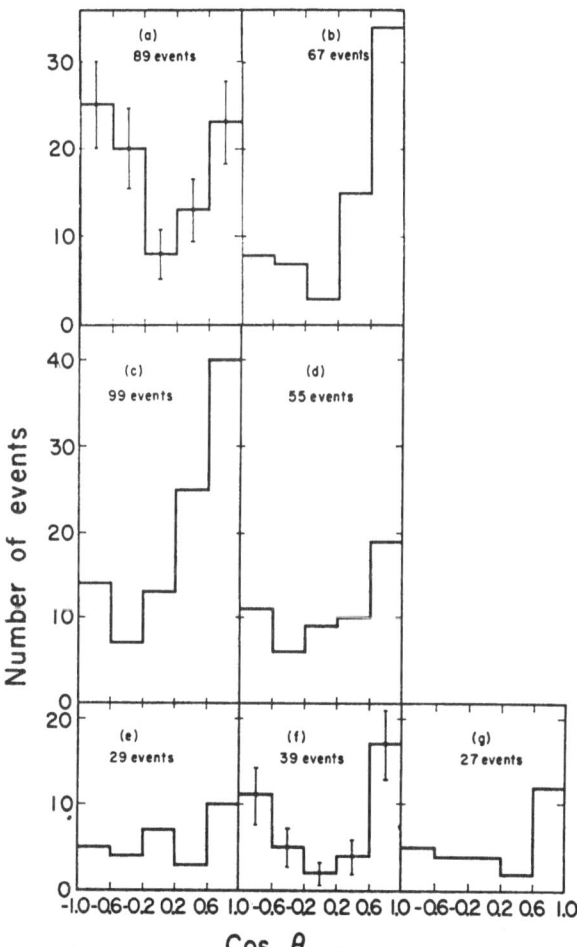

Fig. 19.2. Distributions in $\cos\theta_{\pi\pi}$, the angle between the incoming and outgoing π^- in the barycentric system of the final pions.

(a) $700 \leq M(\pi^-\pi^0) < 850$ MeV
(b) $850 \leq M(\pi^-\pi^0) \lesssim 1000$ MeV
(c) $700 \leq M(\pi^-\pi^+) < 850$ MeV
(d) $850 \leq M(\pi^-\pi^+) < 1000$ MeV

(e) $100 \leq M(\pi^-\pi^+) < 1150$ MeV
(f) $1150 \leq M(\pi^-\pi^+) < 1300$ MeV
(g) $1300 \leq M(\pi^-\pi^+) < 1450$ MeV

19.2(c)]. The f^0 decays approximately symmetrically [Fig. 19.2(f)]. The angular distribution of the f is not isotropic, and spin 0 is therefore ruled out. It is usually assumed that the spin of the f is $J = 2$ because this is the simplest possibility for an $I = 0$ two-pion resonance which does not decay isotropically. However, the decay angular distribution of the f^0 looks almost as much like $\cos^2 \theta$ as like $(3 \cos^2 \theta - 1)^2$. The former angular distribution favors $J = 1$ and the latter favors $J = 2$. It is possible that interference effects and other background can make a spin 2 angular distribution partially simulate a spin 1 distribution. Another possibility, that the f^0 is the neutral component of an $I = 1$, $J^P = 1^-$ meson, has been suggested by FRAZER, PATIL and XUONG (1964). These authors have proposed a model to explain why the charged f mesons, which decay into $\pi^\pm \pi^0$, have not been seen.

Y. LEE et al. (1964) have recently measured the decay angular distribution of the f^0 and found it was necessary to include a $\cos^4 \theta$ term in order to fit the data. This is an additional evidence* for $J = 2$.

B. The B meson

ABOLINS et al. (1963) obtained evidence for a $\pi\omega$ resonance in looking at the reaction

$$\pi^+ + p \rightarrow \pi^+ + p + \omega \tag{19.3}$$

with incident pions of momenta 3.43 and 3.54 GeV/c.

The resonance, called the B meson, has a mass and width given by

$$M = 1220 \text{ MeV}, \qquad \Gamma = 100 \pm 20 \text{ MeV}.$$

Fig. 19.3(a) (ABOLINS 1963) shows a Dalitz plot of reaction (19.3). As can be seen from the figure, the Δ shows up as a prominent peak, while the B is a faint enhancement. In Fig. 19.3(b) is plotted a histogram of the $\pi\omega$ mass spectrum, showing the evidence for the B. In Fig. 19.3(c), events corresponding to Δ formation are subtracted out. This reduces the background, but the peak remains.

KIRZ and MILLER (1963) and BONDAR et al. (1963) have obtained additional evidence for the B in looking at

$$\pi^- + p \rightarrow \pi^- + p + \omega. \tag{19.4}$$

Thus, both a B^+ and a B^- have been seen. The B^0 is expected to decay

$$B^0 \rightarrow \pi^0 + \omega, \qquad \omega \rightarrow \pi^+ + \pi^- + \pi^0.$$

Since the final state has two neutral pions, it will be difficult to observe.

Since the width of the B is large, we assume isospin and G parity are conserved in its decay. This means that the B has $I^G = 1^+$. There is not yet information on the spin and parity of the B. If its spin is odd and its parity is negative, then the B should decay strongly into two pions and two kaons. If the spin and parity are even, strong decay of the B into $\pi\pi$ or $K\overline{K}$ is forbidden.

---

* $J = 1$ is also excluded by the observation of the decay $f^0 \rightarrow 2\pi^0$ (SODICKSON 1964) (Ed.).

The B has a mass 1220 MeV compared to the mass 1250 MeV for the f meson. The widths of these mesons are also approximately the same. There is the remote possibility, mentioned previously, that the f^0 is the neutral component of an $I = 1$ meson. If this is the case, the f^0, B^+ and B^- may be different charge states of the same particles (FRAZER 1963). If so, the quantum numbers are probably $I^G = 1^+$, $J^P = 1^-$, $C = -$.

Other predictions have been made of the spin and parity of the B, including 0^-, 1^+ and 2^-. It is likely that one of the many predictions will turn out to be correct.

A study of the spin and parity of the B meson was recently made by CARMONY et al. (1964), using a test for four-meson resonances proposed by HALPERN (1964). CARMONY et al. found that $J^P = 1^-$ was favored, but that there were insufficient data to exclude other assignments*.

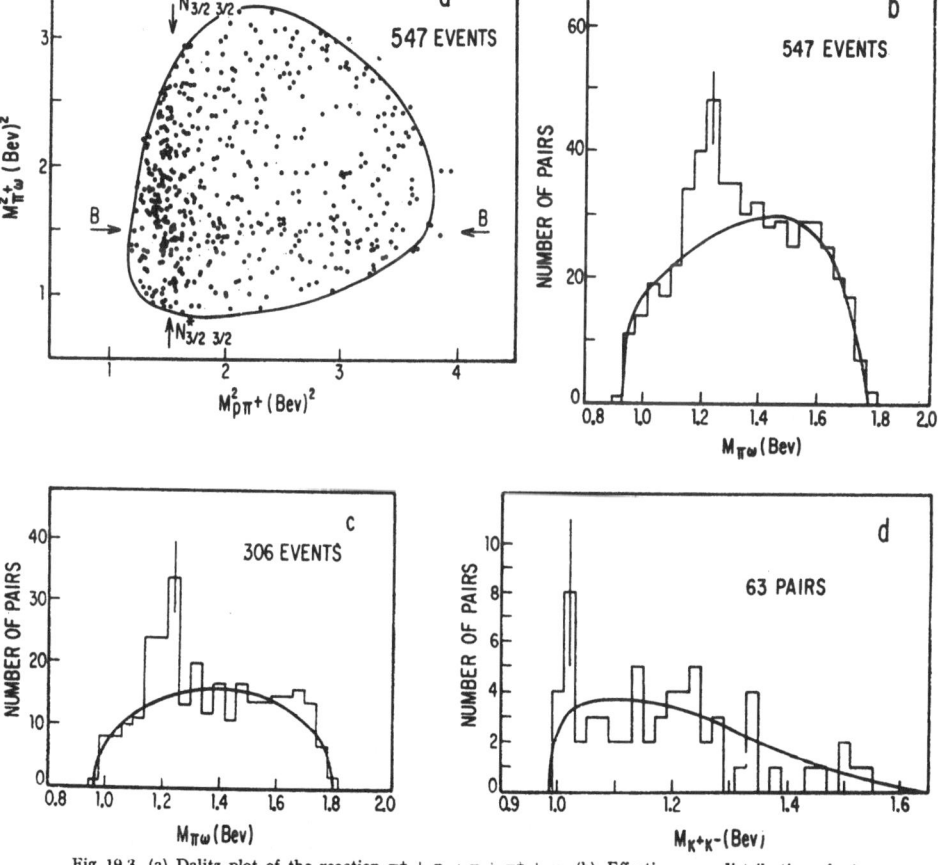

Fig. 19.3. (a) Dalitz plot of the reaction $\pi^+ + p \rightarrow \pi + \pi^+ + \omega$. (b) Effective-mass distribution of $\pi^+\omega$ pairs. (c) Effective-mass distribution of $\pi^+\omega$ pairs when $M(p\pi^+)$ is outside the Δ mass range (1150—1350 MeV). (d) Effective-mass distribution of K^+K^- pairs from the reaction $\pi^+ + p \rightarrow \pi^+ + K^+ + K^- + p$. In (b), (c), and (d), the solid lines are phase space estimates (ABOLINS 1963)

* ADERHOLZ et al. (1964) demonstrate that $I = 0$ (Ed.).

C. The ϰ meson

Evidence for a meson of hypercharge $Y = 1$, called the ϰ, was obtained by ALEXANDER et al. (1962, 1962b) in the reactions

$$\pi^- + p \rightarrow K + \pi + \Sigma$$
$$\rightarrow K + \pi + \Lambda \qquad (19.5)$$

with pions of momenta about 2 GeV/c. The evidence is a small peak in the Kπ mass spectrum at a mass of 730 MeV accompanying a larger peak at the K* mass of 888 MeV.

Further evidence for the existence of the ϰ was obtained by MILLER et al. (1963), who observed a peak in the Kπ mass spectrum with mass and width

$$M = 726 \pm 3 \text{ MeV} \qquad \Gamma = 20 \text{ MeV}$$

in reactions (19.5) at different incident pion momenta. Peaks were observed in both the $\pi^+ K^0$ and $\pi^0 K^+$ modes.

Evidence that the ϰ has isospin $I = {}^1/_2$ comes from the observed branching ratio

$$R = \Gamma(\varkappa \rightarrow \pi^+ K^0)/\Gamma(\varkappa \rightarrow \pi^0 K^+) = 37/18 = 2.0 \ . \qquad (19.6)$$

The predicted ratios are

$$R = 2 \quad \text{for} \quad I = {}^1/_2$$
$$R = {}^1/_2 \quad \text{for} \quad I = {}^3/_2 \qquad (19.7)$$

However, because the ϰ appears as a small peak on a large background, the agreement with the $I = {}^1/_2$ ratio may be fortuitous. Additional evidence that $I = {}^1/_2$ comes from the absence of a peak at the K mass in the $\pi^- K^0$ spectrum. The $\pi^- K^0$ state is pure $I = {}^3/_2$.

MILLER et al. (1963) also looked for a ϰ peak in the $\pi^- K^+$, $\pi^0 K^0$ and $\pi^- K^+$ mass spectra. No peak was observed at 725 MeV, but a peak (or a statistical fluctuation) was seen at $M = 747 \pm 6$ MeV. If the peak is real, it is hard to understand a difference in mass between the charged and neutral ϰ's of 21 MeV.

WOJCICKI et al. (1963) have seen some evidence for the ϰ in the reaction

$$K^- + p \rightarrow \bar{K}^0 + \pi^- + p \ . \qquad (19.8)$$

A small but statistically significant peak at the ϰ mass was seen. The cross section for producing ϰ is only about 1 or 2% of the cross section for producing the K* (888) at K$^-$ momenta between 1.0 and 1.7 GeV/c.

Since thus far, only a few ϰ's have been produced, no evidence has been obtained regarding its spin or parity except that parity conservation in the decay implies that $P = (-1)^J$. It is also possible that the ϰ may not be a meson, but some other effect in the Kπ system.

D. Peaking in the $K\pi\pi$ system

WANGLER et al. (1964) have seen evidence for a peak in the $K\pi\pi$ mass spectrum at a mass and width

$$M = 1170 \text{ MeV}, \qquad \Gamma \leq 50 \text{ MeV} .$$

The peak was observed in the reactions

$$\begin{aligned} \pi^- + p &\to \Lambda + K + \pi + \pi \\ &\to \Sigma + K + \pi + \pi \end{aligned} \qquad (19.9)$$

with incident pions of momentum 3 GeV/c. Since only about 25 events were seen in the peak above a background of 15, nothing is yet known about the quantum numbers. Some events were consistent with the reaction

$$\pi^- + p \to Y + K^* + \pi , \qquad (19.10)$$

but it is not known whether the $K^*\pi$ is an appreciable decay mode of the meson at 1170 MeV.

Some time before the experiment of WANGLER et al., BELYAKOV et al. (1962) reported seeing a number of peaks in the invariant mass spectra of $K\pi\pi$, $K\pi\pi\pi$ and $\pi\pi\pi\pi$ in reactions with 7 GeV/c π^- mesons in propane. The evidence is not convincing because of the small number of events. BELYAKOV et al. reported a mass of about 1150 MeV for the $K\pi\pi$ peak, a value close to the mass observed by WANGLER et al. However, BELAYAKOV et al. observed a peak in the doubly charged states of $K\pi\pi$, but not in states with zero charge, whereas WANGLER et al. have observed the peak in states with charge zero and one.

ARMENTEROS et al. (1964) have seen evidence for a $K\pi\pi$ resonance with mass and width

$$M = 1230 \text{ MeV}, \qquad \Gamma = 80 \pm 20 \text{ MeV} .$$

The two pions seem to prefer to come off with nearly their maximum allowed kinetic energy in their own c. m. system. This may mean that this resonance is a $K\rho$ bound state. It is not yet clear whether this effect is the same as that seen by WANGLER et al.

E. Peaking in the $\rho\pi$ system

CASON et al. (1964) have looked at the reaction

$$\pi^- + p \to p + \pi^+ + \pi^- + \pi^- \qquad (19.11)$$

with 6 GeV/c incident pions. They found that all events could be classified either as

$$\pi^- + p \to \Delta^{++} + \pi^- + \pi^* \qquad (19.12)$$

or

$$\pi^- + p \to p + \rho^0 + \pi^- . \qquad (19.13)$$

CASON et al. made a Dalitz plot of all events which were consistent with the final state $p\rho^0\pi^-$, and observed a peaking in the $\rho\pi$ mass spectrum at low relative momenta of the ρ and π. The peaking in the mass spectrum has a width of over 200 MeV, and is asymmetrical, falling more slowly on the high mass side of the spectrum. The effect appears to be consistent with the assumption that there is a low-energy s-wave $\rho\pi$ interaction of the type that can be described by a scattering length. Other interpretations have not been ruled out, however. The analysis of the experiment is complicated by the fact that the ρ itself has a width of over 100 MeV. If the interpretation is correct, the $\rho\pi$ interaction is in a state $J^P = 1^+$.

GOLDHABER et al. (1964) have seen this reaction in $\pi^+\mathrm{p}$ collisions. They conclude that if the $\rho\pi$ interaction is a resonance, its mass and width are given by

$$M = 1200\ \mathrm{MeV}, \qquad \Gamma = 350\ \mathrm{MeV}\,.$$

F. An effect at mass 960 MeV

GOLDBERG et al. (1964) have looked at the reaction

$$\mathrm{K}^- + \mathrm{p} \to \Lambda + \text{neutrals}$$

with 2.3 GeV/c incident K^- mesons. These authors obtained a spectrum of the total effective mass of the neutral particles emitted in the reaction (the so-called "missing mass" spectrum). Peaks were seen at the mass of the π^0, η, and φ. In addition, a peak of mass about 960 MeV was seen. The significance of this peak is not yet understood[*].

G. A possible $\mathrm{K}\bar{\mathrm{K}}\pi$ resonance

ARMENTEROS et al. (1963) have seen evidence for a $\mathrm{K}\bar{\mathrm{K}}\pi$ resonance at 1410 MeV in an antiproton-proton annihilation experiment. The enhancement was seen only in the $I = 0$ isopin state. The decay observed was $\mathrm{K}_1\mathrm{K}_1\pi^0$. If $I = 0$ is confirmed, this decay mode implies $G = +$.

20. Effects in the $\pi\pi$ and $\mathrm{K}\bar{\mathrm{K}}$ systems

A. The ABC Effect

ABASHIAN, BOOTH and CROWE (1960) measured the recoil He^3 and H^3 momentum spectra from the reactions

$$
\begin{aligned}
\mathrm{p} + \mathrm{d} &\to \mathrm{He}^3 + \pi^0,\ \mathrm{H}^3 + \pi^+ \\
&\to \mathrm{He}^3 + \pi^+ + \pi^-,\ \mathrm{He}^3 + \pi^0 + \pi^0,\ \mathrm{He}^3 + \mathrm{X}^0 \qquad (20.1) \\
&\to \mathrm{H}^3 + \pi^+ + \pi^0,\ \mathrm{H}^3 + \mathrm{X}^+
\end{aligned}
$$

---

[*] The same peak has been observed by KALBFLEISCH et al. (1964), who find $M = 959 \pm 2$ MeV, $\Gamma \leq 12$ MeV, $T = 0$ or 1 and the decay $\pi^+ + \pi^- + \eta$, but not into 2, 3, 4 or 6 pions (Ed.).

at incident proton energies of 624 to 743 MeV. The X^0 is any particle of mass $M_X < 2.8\, M_\pi$. These authors obtained a peak in the He³ spectrum corresponding to an enhancement of whatever else was produced in the reaction at an invariant mass and width

$$M = 310 \text{ MeV}, \qquad \Gamma \approx 25 \text{ MeV}.$$

This enhancement is usually called the ABC. Since the initial state is pure $I = \frac{1}{2}$, the ABC can have isospin either 0 or 1, and the branching ratio into H³ and He³ is a unique function of the isospin.

For events in the peak

$$\begin{aligned} \text{H}^3/\text{He}^3 &= 2 \quad \text{if} \quad I = 1 \\ \text{H}^3/\text{He}^3 &= 0 \quad \text{if} \quad I = 0. \end{aligned} \tag{20.2}$$

The absence of a corresponding bump in the H³ spectrum thus shows that $I = 0$ (BOOTH 1961).

The experiments are described in detail and an analysis is attempted in a series of papers (ABASHIAN 1963, BOOTH 1963). The authors reached the following conclusions:

1. The effect is most likely an enhancement in the channel for production of two pions in an s state. This conclusion is based primarily on the width and low energy of the effect.

2. The effect does not behave kinematically as a particle. This conclusion is based on a "slight but perhaps significant shift" of the mass when the momentum transfer is varied.

3. A fit to the experimental He³ momentum distribution can be obtained by assuming a strong s-wave $\pi\pi$ interaction. In Fig. 20.1 are shown the data and several fits to it. A pion-pion scattering length $(2 \pm 1)/M_\pi$ is needed to obtain a fit.

4. A resonance-type $\pi\pi$ interaction cannot be excluded by the data.

The effect of the ABC was possibly seen by RICHTER (1962) in a proton recoil experiment in the reaction

$$\gamma + \text{p} \rightarrow \text{ABC} + \text{p}. \tag{20.3}$$

RICHTER found a mass $M = 322 \pm 8$ and a width $\Gamma < 20$ MeV, consistent with a zero true width. He was unable to fit his data by assuming an s-wave pion-pion interaction.

The ABC may have also been observed at Frascati (ODIAN 1962) in a photoproduction experiment. In this experiment two charged particles possibly associated with the ABC were seen in a spark chamber. Assuming these to be pions, there are restrictions on the spin and parity of the ABC: $J^P = 0^+, 1^-, 2^+$, etc. This is an unpublished preliminary experiment.

If the ABC effect is really a property of the pion-pion system alone, it should also be seen in pion production experiments in pion-nucleon collisions. A number of attempts have been made to look for the ABC in such collisions, without definite success. A summary of such attempts is given by PUPPI (1963). More information about the ABC is needed before we can decide just what is the nature of this effect *.

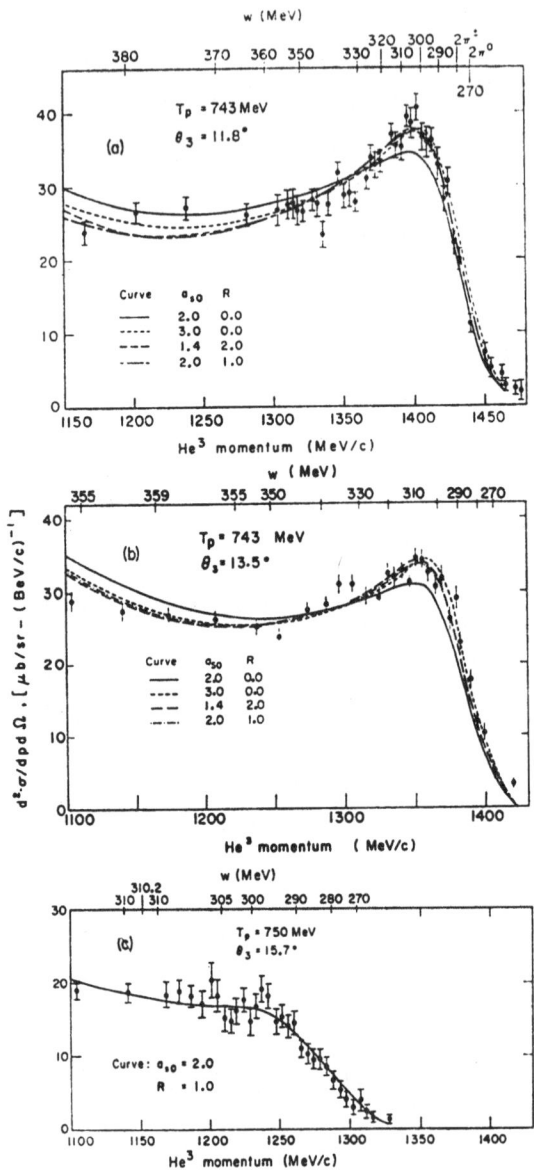

Fig. 20.1. Fits to the experimental data from the reaction p + d→He³ + X⁰ where X⁰ is presumably a two-pion enhancement at 310 MeV. The He³ momentum spectra are fitted assuming a ππ scattering length a_{s0} between 1.4 and 3.0 in units of $1/M_\pi$. Bose symmetry for the pions is taken account, assuming a symmetrization radius R between 0 and 2 in the same units (BOOTH 1963)

---

* See also anomaly in p + n → d + 2π found by HOMER et al. (1964) (Ed.).

B. The pion-pion interaction at 390 MeV

Some inconclusive evidence exists that an s-wave pion-pion interaction exists at an energy somewhat higher than 310 MeV. We briefly
review the evidence.

1. The Dalitz plots of the pions from the decays of the η and K
mesons show that the decay matrix elements do not follow phase space.
Fig. 20.2 (CRAWFORD 1963a) shows the kinetic energy spectrum of the

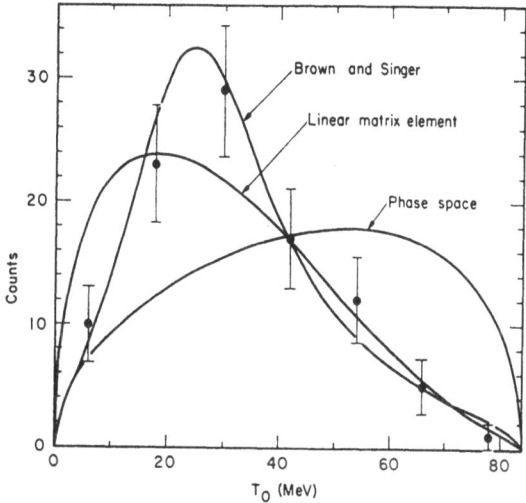

Fig. 20.2. Spectrum of the π^0 kinetic energy from $\eta \to \pi^+ + \pi^- + \pi^0$. The experimental points are from the
experiment of CRAWFORD et al. (1963a). The three solid curves correspond to phase space, the linear matrix
element theories (GELL-MANN 1957a, WALI 1962, BARTON 1962, BÉG 1962), and the $I = 0$, $J = 0$ di-pion-
resonance theory of BROWN and SINGER (1962, 1964) (CRAWFORD 1963a)

π^0 meson from η decay, projected from a Dalitz plot. CRAWFORD et al.
were able to obtain a good fit to the data assuming that there exists
an s wave pion-pion resonance at 385 MeV with a width $\Gamma = 50$ MeV,
as suggested by BROWN and SINGER (1962, 1964). The quantum numbers
are $I^G = 0^+$, $J^P = 0^+$. However, a non-resonant pion-pion interaction
with the same quantum numbers may account for the deviation from
phase space.

2. Many attempts to explain the nucleon-nucleon interaction in
terms of one-meson-exchange graphs require a meson of mass about
$350-400$ MeV and quantum numbers $I^G = 0^+$, $J^P = 0^+$. Again, the use
of this meson may be an approximate way of treating a non-resonant
pion-pion interaction.

3. In an experiment with incident pions of 4.7 GeV/c, SAMIOS et al.
(1962) observed peaks in the $\pi^+\pi^-$ spectrum at masses of 395 MeV and
520 MeV. The 395 MeV peak might be additional evidence for a pion-
pion interaction at this mass. The two peaks have not been established
to be resonances, however.

4. Further evidence of a peak in the two-pion mass spectrum at about 400 MeV was obtained by CALDWELL et al. (1964) in an experiment with high energy (12 and 18 GeV/c) incident pions. This is only a preliminary result, based on one-fourth of the data.

C. The ζ Effect

Weak evidence for a peak in the dipion mass spectrum at a mass about 570 MeV was obtained a long time ago (BARLOUTAUD 1962, ZORN 1962) on the time scale of the discovery of the highly unstable mesons, but confirmatory experiments have not been forthcoming. The evidence of SECHI-ZORN came from a two-pion mass correlation in the reaction

$$p + p \rightarrow d + \pi^+ + \pi^0 . \tag{20.4}$$

Since the two-proton state is pure $I = 1$ and the deuteron is $I = 0$, the ζ must be $I = 1$.

Since the ζ decays into two pions, its spin and parity are $J^P = 0^+$, 1^-, 2^+, 3^-, etc. If it decays via strong interactions then it has odd spin, $P = -$, $G = +$, since these are the possible quantum numbers of a two-pion state with $I = 1$. Then, unless its spin is 3 or higher, it has the same quantum numbers as the ρ meson ($J^{PG} = 1^{-+}$) and should be produced in peripheral collisions like the ρ.

But the ζ is not produced in pion-nucleon collisions with a cross section at all comparable to that of the ρ. Thus, if the ζ is a particle, its quantum numbers are probably different from that of the ρ. A number of possible quantum numbers have been suggested for the ζ on theoretical grounds, among them the quantum numbers $I^G = 1^-$, $J^P = 0^+$. With these quantum numbers, the ζ has the same quantum numbers as a five-pion state, and its decay into pions is an electromagnetic process.

Another possibility is that the ζ is not a meson but a more complicated phenomenon. KENNEY and VITTITOE (1963) and others have observed that in the reactions

$$\pi^- + p \rightarrow n + \pi^- + \pi^+$$
$$\rightarrow p + \pi^- + \pi^0$$

at incident energies below 1 GeV, the mass spectrum of the two pions tends to peak near the high energy kinematic limit. In a number of experiments this peaking occurs near the ζ mass. However, the effect is not a meson, as the peak does not have a Breit-Wigner shape and the mass depends on the incident energy. A number of interpretations of this effect have been given in terms of pion-pion and pion-nucleon interactions, but none of the explanations is completely convincing. See KENNEY and VITTITOE (1963) for details and additional references.

D. The K$\overline{\text{K}}$ system

Evidence exists for two separate effects in the K$\overline{\text{K}}$ system at a mass about 1020 MeV. Two separately resolved peaks have not been seen at this mass, but a peak has shown up in the decay mode $K_1 K_1$ in two experiments (ERWIN 1962, ALEXANDER 1962) and in the mode $K_1 K_2$ in several others. Since the $K_1 K_1$ and $K_1 K_2$ are eigenfunctions of CP with opposite eigenvalues for the same value of the total angular momentum, the $K_1 K_1$ and $K_1 K_2$ cannot arise from the decay of a single meson without violating CP invariance. The $K_1 K_2$ peak arises from the decay of the φ meson. The $K_1 K_2$ peak may arise from the decay of another meson or from an s-waveonteraction of the K$\overline{\text{K}}$ system.

If CP is conserved, then a single meson can decay into both $K_1 K_1$ and $K_1 K_2$ only if it is not an eigenstate of CP and has a distinct antiparticle. Then one linear combination of the meson and its antiparticle can be formed which is even under CP and another which is odd. In such a case, we can confine our attention to the two linear combinations, which are distinct mesons. Furthermore, since the evidence indicates that the particles decaying into $K_1 K_1$ and $K_1 K_2$ are produced in different reactions and have different widths, there is at present no reason (apart from a similarity in masses) to identify them as a single meson. Both the φ and the meson (or K$\overline{\text{K}}$ interaction) decaying into $K_1 K_1$ have the decay mode $K^+ K^-$. This means that a $K^+ K^-$ pair with a mass about 1020 MeV can arise from two separate effects (or be part of the general background).

We briefly review the relevant experiments concerning the $K_1 K_1$. ERWIN et al. (1962) considered K$\overline{\text{K}}$ pairs produced in the reactions

$$\pi^- + p \rightarrow K^0 + \overline{K}^0 + n$$

$$\rightarrow K^- + K^0 + p$$

$$\rightarrow K^+ + K^- + n$$

by 1.89 and 2.10 Gev/c pions. A mass spectrum of the $K^- K^0$ pairs followed phase space within experimental error, but the $K^0 \overline{K}^0$ pairs appeared somewhat concentrated at low mass values. All $K^0 \overline{K}^0$ pairs were observed as $K_1 K_1$ since the mass spectrum of $K_1 K_2$ pairs could not be obtained in the experiment. Not enough $K^+ K^-$ pairs were observed for a mass spectrum of this mode to be meaningful. The results of ERWIN et al. (1962), while not statistically convincing by themselves, have been supported by a similar experiment of ALEXANDER et al. (1962) at incident pion momenta less than 2.3 Gev/c.

We make use of the selection rules given previously to obtain information on the quantum numbers of the $K_1 K_1$. First, note that $J(K_1 K_1)$ is even, assuming the $K_1 K_1$ system obeys Bose statistics. Assuming P and C are separately conserved in the decay; i.e., that the decay is into the mode $K^0 \overline{K}^0$ with the $K^0 \overline{K}^0$ subsequently decaying into $K_1 K_1$, the quantum numbers of the effect are

$$J^{PG} = 0^{++}, 2^{++} \cdots \quad \text{if} \quad I = 0 ,$$
$$J^{PG} = 0^{+-}, 2^{+-} \cdots \quad \text{if} \quad I = 1 ,$$

since $G = (-1)^I C$. The evidence for $I = 0$ is that the mass spectrum of $K^- K^0$ follows phase space in the experiments of ERWIN et al. and ALEXANDER et al.

Further evidence for $I = 0$, $G = +$ is that the nucleon tends to go backward in the c.m. system, and no correlations have been seen in the production and decay planes of the $K_1 K_1$. These results are consistent with a model in which the $K_1 K_1$ is produced in peripheral collisions via one pion exchange. Such a model requires I even, $G = +$ if J is even. In the spirit of this model, an estimate of the cross section $\sigma (\pi \pi \to K \overline{K})$ can be obtained using the Chew-Low formula. See ERWIN et al. (1962) and ALEXANDER et al. (1962) for details.

ALEXANDER et al. have been able to fit the $K_1 K_1$ mass spectrum assuming an s-wave $K \overline{K}$ interaction in a zero effective range approximation. They find that the scattering amplitude is well represented by a complex scattering length. If this interpretation is correct, the $K_1 K_1$ has $I = 0$, $J^{PG} = 0^{++}$. These authors also point out that if $J = 2^{++}$, a much more rapid decay into two pions would be expected, whereas a two-pion resonance at $M = 1020$ MeV has not been seen. However, such an argument depends on dynamical assumptions, and is not good evidence against 2^{++}. For example, a broad two-pion resonance centered at an energy below two K masses might not have been seen. Another (rather remote) possibility is that $I = 1$, $G = -$, an assignment which forbids decay into two pions. The behavior of the $K^- K^0$ mass spectrum is evidence against this possibilty.

Fig. 19.3 (d) (ABOLINS 1963) shows a mass spectrum of $K^+ K^-$ pairs from the reaction

$$\pi^+ + p \to \pi^+ + p + K^+ + K^- .$$

There appears to be some evidence for an enhancement at 1020 MeV. There are very few events, but if the narrow width is real, what is seen is the φ rather than a $K \overline{K}$ s-wave interaction.

It is significant that the $K_1 K_1$ interaction is observed chiefly in interactions with an incident pion, while the φ is observed principally in K^- interactions. This fact may be partly understood in terms of a one-meson-exchange model of production. The $K_1 K_1$ can be produced in a collision with an incident pion with the exchange of a single pion. The φ, however, cannot. The φ seems to have the right quantum numbers to be produced in πN collisions via ρ exchange. However, as we have seen, the $\pi \rho \varphi$ coupling is very small, as evidenced by the low rate of the decay $\varphi \to \rho + \pi$. On the other hand, both the φ and the $K_1 K_1$ can be reached in $K^- N$ collisions via single K exchange. The fact that the φ is produced with greater probability may be an indication that the $K_1 K_1$ is not a resonant state.

21. Excited states of the nucleon

A. Pion-nucleon total cross sections

Many experiments have been performed with π^+ and π^- mesons incident on protons. Other experiments have been done with π^+ and π^- incident on deuterons to obtain information on the pion-neutron cross sections. These latter experiments have generally tended to confirm charge independence and, as a byproduct, to yield information about the deuteron. We shall consider only experiments with proton targets.

A pion and a nucleon can combine in a state of total isospin $I = {}^3/_2$ or $^1/_2$. The π^-p system is a coherent mixture of these two states. Let the total collision cross sections in the $I = {}^3/_2$ and $I = {}^1/_2$ states be σ_3 and σ_1. These cross sections are given in terms of the π^+p and π^-p total cross sections σ_+ and σ_- by the following formulas

$$\sigma_3 = \sigma_+$$
$$\sigma_1 = {}^1/_2 \, (3\sigma_- - \sigma_+)$$
(21.1)

Knowledge of the pure isospin cross sections σ_3 and σ_1 is useful to obtain the isospins of resonances. In Fig. 21.1, σ_3 and σ_1 are plotted as a func-

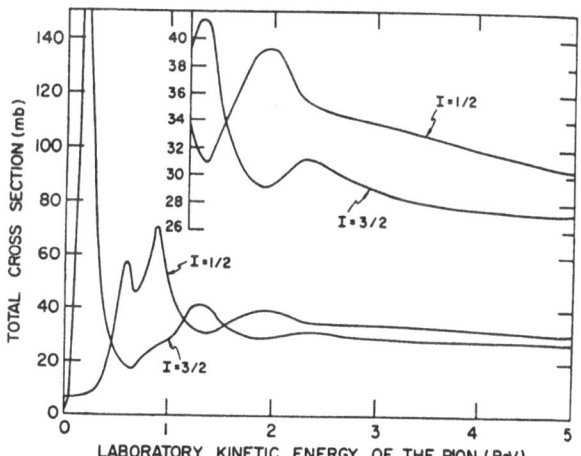

Fig. 21.1. Pion-nucleon total cross sections in the two isospin states. For pion kinetic energies greater than 1.2 GeV, the cross sections are magnified ten times (DIDDENS 1963)

tion of the pion laboratory kinetic energy (DIDDENS 1963). The $I = {}^3/_2$ cross section shows three distinct peaks and a "shoulder" which may be a fourth peak superimposed on a rapidly rising background cross section. The $I = {}^1/_2$ cross section also exhibits three distinct peaks. The highest energy peaks in both σ_3 and σ_1 are also given on an expanded scale, as they are barely visible on the regular scale. All these peaks are states of hypercharge $Y = 1$.

B. The Δ baryon

The first evidence for the existence of an excited state of the nucleon came from pion-nucleon scattering experiments below 300 MeV pion laboratory kinetic energy. An excellent review of the low energy experiments and their interpretation is given in the book by BETHE and DE HOFFMANN (1955).

The most striking feature of the low energy data is the appearance of peaks in both the $\pi^+ p$ and $\pi^- p$ total cross sections at a pion laboratory kinetic energy of 195 MeV. The peak appears at mass and width

$$M = 1238 \pm 2 \text{ MeV}, \qquad \Gamma = 90 \text{ MeV} .$$

The peak was established to be a resonance in the πN system with isospin $I = \frac{3}{2}$ and spin and parity $J^P = \frac{3}{2}^+$. Various symbols for the peak are the (3,3) resonance, $N_{\frac{3}{2}}^*$, and Δ. We briefly discuss the evidence for these quantum numbers.

Let the pion-nucleon scattering amplitudes in the $I = \frac{3}{2}$ and $I = \frac{1}{2}$ isospin states be a_3 and a_1. Then the π^+ cross section σ_+ and the π^- elastic and charge exchange cross sections σ_e and σ_c defined by

$$\sigma_+ = \sigma(\pi^+ p \to \pi^+ p), \quad \sigma_e = \sigma(\pi^- p \to \pi^- p), \quad \sigma_c(\pi^- p \to \pi^0 n) \quad (21.2)$$

can be written in terms of these amplitudes as follows

$$\begin{aligned}
\sigma_+ &= |a_3|^2 \\
\sigma_e &= \frac{1}{9} |a_3|^2 + \frac{4}{9} |a_1|^2 + \frac{4}{9} \operatorname{Re} a_3^* a_1 \qquad (21.3) \\
\sigma_c &= \frac{2}{9} |a_3|^2 + \frac{2}{9} |a_1|^2 - \frac{4}{9} \operatorname{Re} a_3^* a_1 .
\end{aligned}$$

The interference term cancels in the total π^- cross section $\sigma_- = \sigma_e + \sigma_c$. It follows that the ratio of σ_- to σ_+ is given by

$$\sigma_-/\sigma_+ = \frac{1}{3} + \frac{2}{3} |a_1|^2/|a_3|^2 . \qquad (21.4)$$

But experimentally it is observed that

$$\sigma_-/\sigma_+ = \frac{1}{3} \qquad (21.5)$$

showing that $|a_1|^2 \ll |a_3|^2$. In other words, the $I = \frac{3}{2}$ amplitude is dominant. Consistent with this evidence are the experimental ratios

$$\sigma_+ : \sigma_c : \sigma_e = 9 : 2 : 1 . \qquad (21.6)$$

This shows that

$$\operatorname{Re} a_3^* a_1 \ll |a_3|^2 . \qquad (21.7)$$

The spin is determined from the observed angular distribution

$$I(\Theta) = 1 + 3 \cos^2 \Theta \qquad (21.8)$$

which is characteristic of a pure $J = \frac{3}{2}$ state. This is confirmed by the fact that the observed magnitude of the cross-section at the resonance is given by the unitarity limit

$$\sigma_+ = 8\pi/k^2 = 2\pi(2J + 1)/k^2 \quad \text{with} \quad J = \frac{3}{2} \qquad (21.9)$$

where k is the relative momentum in the c.m. system.

Because of the Minami ambiguity, neither the angular distribution nor the magnitude of the cross section at the resonance gives information about the parity. The evidence that the resonance has positive parity is that the dominant phase shift δ_{33} at low energy goes as k^3, a behavior which indicates a p-wave and therefore odd parity of the orbital wave function. Since the pion has negative parity, the total parity is positive. Detailed phase shift analyses, making use of Coulomb interference effects and using a smooth behavior of the phase shifts as a function of energy, confirm these conclusions.

C. The N (1512)

In Fig. 21.2 is plotted the $\pi^- p$ total cross section in the energy range between 0.4 and 1.6 GeV (DEVLIN 1962). There is clear evidence for peaks at 600 and 900 MeV incident energy. That these peaks occur primarily

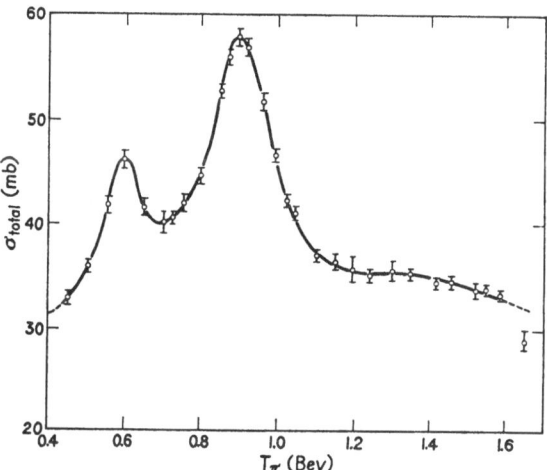

Fig. 21.2. Total cross section for $\pi^- p$ as a function of the laboratory pion kinetic energy in the pion energy range 0.4 to 1.6 GeV (DEVLIN 1962)

in the $I = \frac{1}{2}$ state is shown by the absence of corresponding peaks at these energies in the $\pi^+ p$ cross section of Fig. 21.3 (DEVLIN 1962). (See Fig. 21.1 for the cross sections in pure isospin states.)

We first discuss briefly the 600 MeV peak. This peak, the N (1512) or $N^*_{\frac{1}{2}}$, is at a mass

$$M = 1512 \pm 2 \text{ MeV}$$

and has a width

$$\Gamma = 100 \text{ MeV}$$

which is not too well determined because of the large background.

Additional evidence supporting the $I = \frac{1}{2}$ assignment comes from the photo-pion production reactions

$$\gamma + p \to \pi^+ + n, \qquad \gamma + p \to \pi^0 + p. \qquad (21.10)$$

The predicted branching ratio π^+/π^0 is given by

$$\pi^+/\pi^0 = 2 \quad \text{if} \quad I = \frac{1}{2},$$
$$\pi^+/\pi^0 = \frac{1}{2} \quad \text{if} \quad I = \frac{3}{2}. \qquad (21.11)$$

Experimentally, it is found that

$$\pi^+/\pi^0 \approx 2,$$

supporting $I = \frac{1}{2}$.

The spin and parity of this state are usually taken to be $J^P = \frac{3}{2}^-$. This assignment corresponds to a resonance in pion-nucleon scattering in the $d_{3/2}$ $(L = 2)$ state. The evidence for $J^P = \frac{3}{2}^-$ comes from elastic $\pi^- p$ angular distributions, (CENCE 1963) and photoproduction data. See the analysis of PEIERLS (1960).

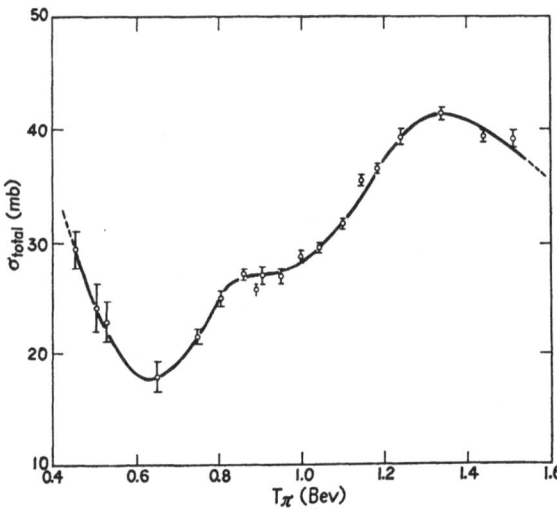

Fig. 21.3. Total cross section for $\pi^+ p$ as a function of the laboratory pion kinetic energy in the pion energy range 0.4 to 1.6 GeV (DEVLIN 1962)

However, the evidence is by no means conclusive that the peak comes from a pion-nucleon resonance in a single angular momentum and parity state. There are two difficulties which plague attempts to analyze the pion-nucleon elastic scattering data at incident energies above 500 MeV.

1. As the energy increases so does the number of phase shifts required in a partial wave analysis.

2. In addition to elastic scattering, there is a rapidly increasing amount of production of an additional pion. This means that the phase shifts used in the analysis must be complex, and that therefore the number of

real parameters is increased by a factor of two. Furthermore, the rapidly increasing inelastic cross section may mean that the peaks in the total cross section are intimately connected with the absorptive part of the scattering amplitude. When absorption is present, the complex scattering amplitude responsible for a resonance moves in a counterclockwise circle as the energy increases. If the absorptive part of the amplitude in the resonant state is sufficiently large, the real part of the corresponding phase shift need not become 90° at the resonance.

3. Another complication comes from the effects of pion-pion interactions. Consider the reaction $\pi + N \to N + \pi + \pi$. If the two final-state pions interact in a state of definite spin, isospin, and parity the $\pi\pi N$ system may be in a mixture of states.

In view of the possibility for all these effects, it is not surprising that it is possible to obtain phase shift analyses of the elastic $\pi^- p$ scattering in which there is not a single dominant phase shift at the 1512 MeV peak (MOYER 1961, CENCE 1963). Nevertheless, it is simplest to regard each peak in the pion-nucleon cross section as the manifestation of a resonance in a single state of total angular momentum and parity. If such is the case with the peak at a mass of 1512 MeV, its most probable quantum numbers are $J^P = {}^3/_2{}^-$.

ROPER (1964) has made a phase shift analysis of pion-nucleon scattering and finds that a $d_{3/_2}$ resonance occurs at an energy of 1559 MeV, rather than at 1512 MeV where the peak in the pion-nucleon cross section occurs. He also finds evidence for a $p_{1/_2}$ resonance at 1485 MeV. Thus, the peak at 1512 MeV may result from the superposition of two resonances, N (1485) and N (1559) *.

D. The N (1688)

The peak in the $\pi^- p$ total cross section at 900 MeV incident π^- kinetic energy corresponds to a mass and width

$$M = 1688 \pm 3 \text{ MeV}, \qquad \Gamma = 100 \text{ MeV}.$$

Since this peak does not appear in the $\pi^+ p$ cross section, its isospin is $I = {}^1/_2$.

The same remarks that were made about the N (1512) apply to the N (1688) as well. Phase shift analyses require far too many parameters to provide a unique fit to the available data. However, it is consistent with the experimental information to assume that there is a resonance in the $f_{5/_2}$ state of the pion and nucleon; i.e., that the spin and parity of the N (1688) are $J^P = {}^5/_2{}^+$.

Various groups (BERTANZA 1962, EISLER 1961, WOLF 1961) have observed a peak at a mass of about 1700 MeV in the cross-section for the

* See also the work of BAREYRE et al. (1964) (Ed.).

reaction

$$\pi^- + p \rightarrow \Lambda + K^0 \tag{21.12}$$

at an incident pion energy of 900 MeV. While at first sight this might be considered an alternate decay mode of the well-established N (1688), there is some evidence that it is not. FELD and LAYSON (1962) have shown that the data on ΛK production can be fitted with a smooth background plus a $p_{1/2}$ resonance. This analysis is not unique, however. Since several fits to the $\pi^- p$ scattering and photoproduction data indicate that the N (1688) is an $f_{5/2}$ resonance, there may be two different resonances, $1/2^+$ and $5/2^+$ at a mass of about 1700 MeV.

E. The Δ (1920) excited state

There is a rather broad maximum in the $\pi^+ p$ total cross section at an incident pion energy of 1305 MeV (Fig. 21.3). This peak corresponds to an invariant mass and width of the pion-nucleon system of

$$M = 1920 \pm 15 \text{ MeV}, \qquad \Gamma = 200 \text{ MeV}.$$

A series of elastic scattering differential cross section measurements were made at incident pion energies between 500 and 1600 MeV (HELLAND 1962). The observed angular distribution was fitted to a power series in $\cos\Theta$ of the form

$$I(\Theta) = \sum_i a_i \cos^i\Theta$$

where the a_i are parameters. It was found that a_6 increases rapidly as the energy approaches the resonance. This fact is most easily interpreted if the resonance has a spin $J = 7/2$, but of course this interpretation is not unique. The parity of the state is unknown, but is often assumed to be positive for theoretical reasons.

Some evidence has also been found by ERWIN et al. (1962a) for a ΣK resonance at a mass of 1920 MeV. This may be an alternate decay mode of the Δ (1920), although its width appears somewhat narrower. If the ΣK is a manifestation of the Δ (1920), then it has $I = 3/2$. KUZNETSOV et al. (1962) may have seen a peak in the ΛK^0 system at a mass of 1840 MeV. If this is interpreted as a $\Sigma^0 K^0$ system followed by $\Sigma^0 \rightarrow \rightarrow \Lambda + \gamma$, then $M_{\Sigma K} = 1920$ MeV. No ΛK resonance has been seen at this mass, a fact which is consistent with $I = 3/2$.

F. The N (2190)

DIDDENS et al. (1963) measured the $\pi^- p$ total cross section in a very accurate transmission experiment at incident pion momenta between

1.5 and 4.5 GeV/c. (See Fig. 21.4.) The authors observed a peak of 2 mb with a mass and width

$$M = 2190 \pm 20 \text{ MeV}, \qquad \Gamma = 200 \text{ MeV}.$$

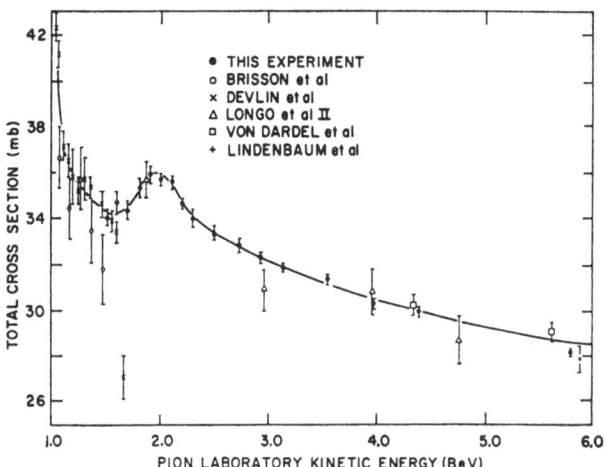

Fig. 21.4. Total π^-p cross section in the energy range 1 to 6 GeV (DIDDENS 1963)

Although the height of the peak is less than 10% of the 30 mb background, the errors are small enough to make the peak convincing. The same authors measured the π^+p total cross section and found no peak at this mass. This shows that if the peak arises from a pure state, it is $I = \frac{1}{2}$. Nothing is known about the spin or parity.

G. The Δ (2360)

DIDDENS et al. (1963) measured the π^+p total cross section in the same momentum range as the π^-p cross section. They observed a 2 mb peak (Fig. 21.5) at a π^+ incident energy of 2.37 GeV. The mass and width of this peak are

$$M = 2360 \pm 25 \text{ MeV}, \qquad \Gamma = 200 \text{ MeV}.$$

The isospin of the state is $I = \frac{3}{2}$, and the spin and parity are not known.

Since the N (2190) and Δ (2360) are produced as small peaks above a large background, it will be difficult to measure their spins and parities, if indeed they are resonances. It may be helpful to study the various inelastic channels that contribute to the cross sections at these energies. The ratio of peak to background may be greater in some particular channel than in the total cross section.

No evidence for any structure in either the π^+p or π^-p cross section at incident energies above 2.37 GeV has been obtained, as can be seen

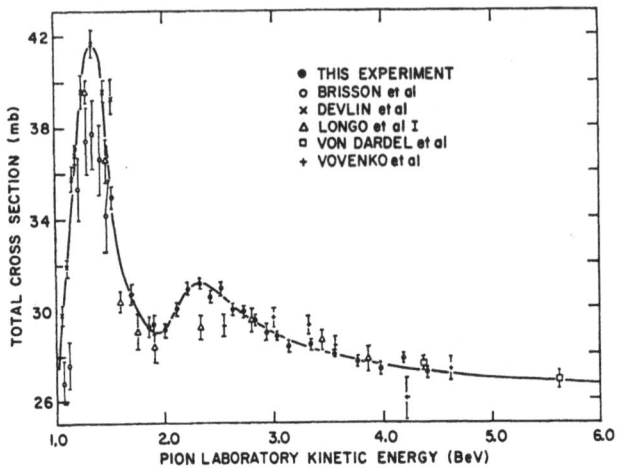

Fig. 21.5. Total π^+p cross section in the energy range 1 to 6 GeV (DIDDENS 1963)

from Fig. 21.6 (LINDENBAUM 1961). In view of the fact that the N (2190) and Δ (2360) are small peaks on large backgrounds, extremely precise measurements will have to be made to detect higher energy resonances in the total pion-nucleon cross sections, provided such resonances exist.

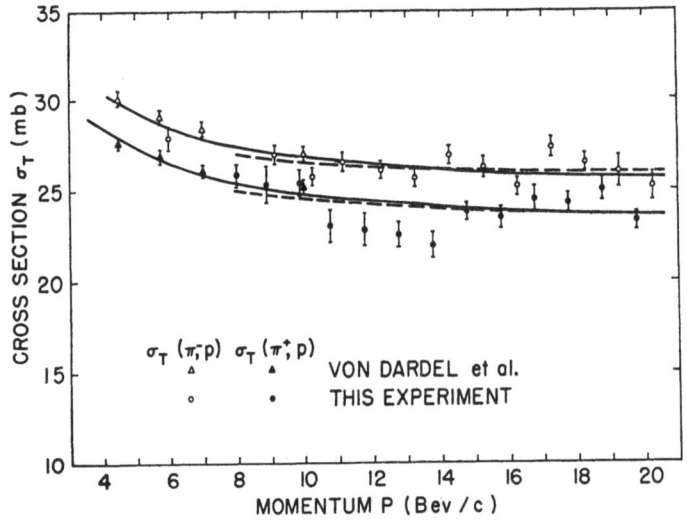

Fig. 21.6. Total π^+p and π^-p cross sections at high energy as a function of the pion laboratory momentum (LINDENBAUM 1961)

H. The "shoulder"

There is a "shoulder" in the π^+p total cross section at a mass of 1650 MeV (Fig. 21.3). Some people have interpreted this effect as a resonance superimposed on a rapidly rising background. At present, such an interpretation is not at all necessary to fit the data, and there is also no compelling theoretical reason for making this interpretation.

I. Systematics

Many attempts have been made to understand the excited nucleon states in terms of dynamical models. See, for example, CARRUTHERS (1964). Here we simply give some empirical systematics.

HELLAND (quoted by ROSENFELD 1963a) has suggested an empirical selection rule relating the isospin I and the spin J of the nucleon and each of its excited states to the orbital angular momentum L of the state. (The orbital angular momentum is that of the pion, considering the excited nucleon states as pion-nucleon resonances and the nucleon as a pion-nucleon bound state.) HELLAND's rule, recently emphasized by KYCIA and RILEY (1963), is

$$J - L = I - 1 . \qquad (21.13)$$

Since the parities of a number of the excited nucleon states are not known, the rule is based on two cases, N and Δ, and on two probable cases, N (1512) and N (1688).

Using this rule, plus other guess work, KYCIA and RILEY have classified the excited states as shown in Table 21.1. The π^+p "shoulder" is optimistically considered as a resonance in this classification.

Table 21.1. Parameters of the nucleon and its excited states. A question mark after the spin or parity of an excited state means that the assignment is in doubt. The J^P assignments of N [2190], Δ (1650), and Δ (2360) are based on an empirical rule plus guesswork (KYCIA 1963)

Isospin	New symbol	Old symbol	Mass MeV	Width MeV	Pion laboratory energy MeV	Spin and parity	Orbital angular momentum of pion
$1/2$	N	N	939	stable	bound state	$1/2^+$	1
$1/2$	N (1512)	$N^*_{1/2}$	1512	100	600	$3/2^-$?	2 ?
$1/2$	N (1688)	$N^{**}_{1/2}$	1688	100	900	$5/2^+$?	3 ?
$1/2$	N (2190)	$N^{***}_{1/2}$	2190	200	1950	$7/2^-$?	4 ?
$3/2$	Δ	$N^*_{3/2}$	1238	90	200	$3/2^+$	1
$3/2$	Δ (1650) ?	"shoulder"	1650	150	860	$5/2^-$?	2 ?
$3/2$	Δ (1920)	$N^{**}_{3/2}$	1920	200	1350	$7/2^+$?	3 ?
$3/2$	Δ (2360)	$N^{***}_{3/2}$	2360	200	2370	$9/2^-$?	4 ?

22. The Σ (1385) hyperon

A. Mass, width and isospin

The first pion-hyperon resonance to be discovered was the Σ (1385) or the Y_1^*. It appeared as a $\Lambda\pi$ resonance which was seen by ALSTON et al. (1960) in the reaction

$$K^- + p \to \Lambda + \pi^+ + \pi^- . \qquad (22.1)$$

The mass and width of this excited state are

$$M = 1385 \pm 3 \text{ MeV}, \qquad \Gamma = 50 \pm 10 \text{ MeV} .$$

The isospin is $I = 1$, as can be seen from the decay mode $\Sigma^* \to \Lambda + \pi$.

Fig. 22.1 shows a Dalitz plot of the events, together with projections on axes corresponding to the kinetic energy of the π^+ and π^-. The histogram showing the number of events versus the kinetic energy of the π^- has two peaks. The one at a π^- kinetic energy of 280 MeV is

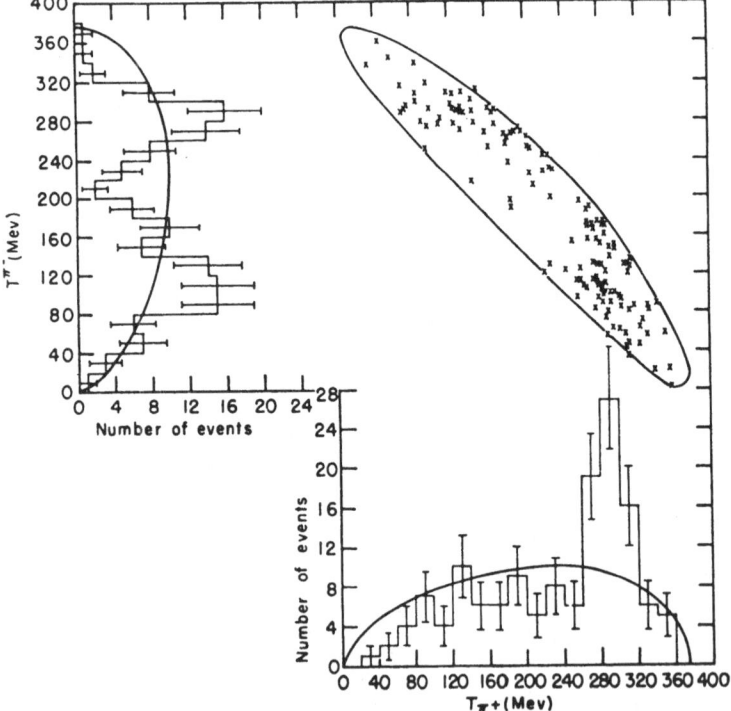

Fig. 22.1. Energy distribution of the two pions from the reaction $K^- + p \to \Lambda + \pi^+ + \pi^-$. Each event is plotted only once on the Dalitz plot, which should be uniformly populated if phase space dominated the reaction. The two energy histograms are merely one-dimensional projections of the two-dimensional plot, and each event is represented once on each histogram. The solid lines superimposed over the histograms are the phase-space curves (ALSTON 1960)

evidence for a $\pi^+ \Lambda$ resonance at an invariant mass of 1385 MeV. Note that there is another peak in this spectrum at a π^- kinetic energy of 100 MeV. Inspection of the Dalitz plot indicates that, in part at least, this second peak is a manifestation of the fact that the π^+ also likes to be emitted at a unique energy because of a $\pi^- \Lambda$ resonance at a mass of 1385 MeV. The $\pi^+ \Lambda$ and $\pi^- \Lambda$ resonances at the same energy are simply different charge states of the Σ (1385). Thus the second peak is not necessarily evidence for another $\Lambda \pi$ resonance. However, careful study of the region of the peak at a π^- kinetic energy of 100 MeV has in fact given indication for a second $\Lambda \pi$ resonance of mass $M = 1660$ MeV. However, most of the events in the second peak are manifestation of the Σ (1385) rather than of the new resonance.

B. Spin

ALSTON et al. (1960) examined the angular distribution of the decays of the Σ^*, and obtained results consistent with isotropy. Spin higher than $1/_2$ was not ruled out, however.

Good evidence concerning the spin of the Σ^* comes from an experiment by ELY et al. (1961) who looked at Σ^* hyperons produced in $K^- + p \rightarrow \Lambda + \pi^+ + \pi^-$ at 1.11 GeV/c. An Adair analysis with $\cos \Theta > 0.9$ yielded results consistent with either $J = 1/_2$ or $3/_2$. However, an examination of the decay angular distribution of the Σ^* with respect to the perpendicular to the production plane yielded the result

$$I(\theta) = 1 + a \cos \theta + b \cos^2 \theta ,$$

$$a = 0.07 \pm 0.24, \qquad b = 1.5 \pm 0.4 .$$

This is almost four standard deviations from isotropy. For the decay to be anisotropic (if $J > 1/_2$), the Σ^* should be produced polarized. The degree of polarization depends on the angle of production. For this analysis only events with $|\cos \Theta| < 0.5$ were used, there being 143 such events out of a total sample of 378. The small value of the asymmetry coefficient a in the analysis is consistent with the Σ^* decaying as a free particle, but by no means proves this. A peculiarity of the data of ELY et al. is that $M_{\Sigma^*} = 1376 \pm 3$ MeV, a result below the accepted value. This may indicate that interference effects actually are present. SHAFER et al. (1963), in an experiment with more than twice as many events as used by ELY et al., found a smaller coefficient b of $\cos^2 \theta$

$$b = 0.69 \pm 0.22 .$$

Since this experiment was at a K^- incident momentum of 1.22 GeV/c, the polarization of the Σ^* may be varying rapidly in the region between incident momenta of 1.11 and 1.22 GeV/c.

The data of ELY et al. and SHAFER et al. on the decay angular distribution are shown in Fig. 22.2. These data, taken together, are very good evidence that the spin of the Σ^* is greater than $^1/_2$. Furthermore, since in no experiment has a $\cos^4\theta$ term in the angular distribution been seen, the spin is very likely $J = ^3/_2$.

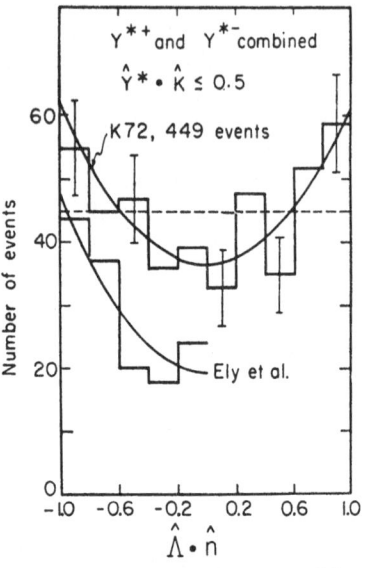

Fig. 22.2. The decay angular distribution of the Σ (1385) (Y* in the Figure) with respect to the normal to the production plane. The abscissa represents the scalar product of a unit vector along the Λ momentum with the unit vector along the normal in the Σ^* rest frame (SHAFER 1963)

C. Parity

In an experiment by COLLEY et al. (1962), Σ^* hyperons were produced in the reaction

$$\pi^- + p \to \Lambda + \pi^- + K^+$$

at an incident π momentum of 2.0 GeV/c. The angular distribution of the decay of the Σ^* with respect to the normal to the production plane was found to yield the following values of the a and b coefficients

$$a = 0.29 \pm 0.29, \qquad b = 1.29 \pm 0.78 .$$

Since a is small, interference effects may not be important. The large value of b indicates $J > ^1/_2$, but the error is such that $J = ^1/_2$ is not excluded.

However, this experiment is more interesting in what it has to say about the Σ^* parity. Because of the Minami ambiguity, the parity of the Σ^* cannot be determined from the angular distributions of its decay into $\Lambda + \pi$. However, a measurement of the polarization of the Λ can give information about the Σ^* parity. Because the Λ decays with non-conservation of parity, the asymmetry of the Λ decay gives information about the Λ polarization which in turn gives information about the Σ^* polarization. COLLEY et al. measured the asymmetry in the Λ decay and found an average value of the polarization $|\overline{P}_\Lambda| = 0.82 \pm 0.27$. For the observed decay angular distribution $I(\theta)$, the maximum calculated value of $\overline{P}_\Lambda$ is $|\overline{P}_\Lambda|_{\max} = 0.47 \pm 0.09$ if the Σ^* is a $p_{3/2}$ state of $\Lambda\pi$; i.e., if $P_{\Sigma^*} = +$. Likewise, $|\overline{P}_\Lambda|_{\max} = 0.28 \pm 0.05$ if the Σ^* is a $d_{3/2}$ state. Thus, if $J = ^3/_2$, as the evidence indicates, the experiment favors positive Σ^* parity. Evidence from the experiment of SHAFER et al. (1963) supports this conclusion. For the detailed analysis, consult the original paper.

With the quantum numbers $J^P = ^3/_2{}^+$, the Σ^* is a $\pi\Lambda$ scattering resonance which is analogous to the $J = ^3/_2{}^+$ πN scattering resonance (the Δ).

D. Decay

The dominant decay mode of the Σ^* is

$$\Sigma^* \to \Lambda + \pi .$$

A puzzling feature is the small branching ratio into the mode $\Sigma + \pi$. Experimentally, it is observed that

$$R = \Gamma(\Sigma^* \to \Sigma\pi)/\Gamma(\Sigma^* \to \Lambda\pi) = 0.02 \pm 0.02 .$$

The $\Sigma\pi$ decay mode is expected to be somewhat inhibited by the smaller available phase space. To estimate the effects of the difference in the $\Sigma\pi$ and $\Lambda\pi$ relative momenta, assume a simple model in which the decay rate Γ into a particular 2-body modes goes as

$$\Gamma \sim p\left(\frac{p^2 a^2}{1 + p^2 a^2}\right)^L ,$$

where p is the relative momentum of the two final-state particles, L is the orbital angular momentum, and a is an appropriate interaction radius. Using this model, we obtain for R the value

$$R = 0.25 ,$$

assuming p-wave decay in both the $\pi\Lambda$ and $\pi\Sigma$ modes and an interaction radius $a = 1/(2M_\pi)$. As in the cases of other particles, dynamical models have been proposed to explain the discrepancy between the observed result and the phase space calculation. We shall not discuss these models.

23. The Λ (1405) hyperon

A. Mass and width

Evidence for a $\Sigma\pi$ resonance was observed by ALSTON et al. (1961 a) in K^-p reactions at 1.15 GeV/c. The reactions studied were

$$\begin{aligned}
K^- + p &\to \Sigma^\pm + \pi^\mp + \pi^+ + \pi^- \\
&\to \Sigma^0 + \pi^0 + \pi^+ + \pi^- .
\end{aligned} \tag{23.1}$$

Supporting evidence came from an experiment by BASTIEN et al. (1961). Later more conclusive data were obtained by a number of groups (COOPER 1962, ALSTON 1962). The mass and width of this $\Sigma\pi$ resonance are

$$M = 1405 \text{ MeV}, \qquad \Gamma = 50 \text{ MeV} .$$

A much narrower width for the Λ (1405) has been reported by FRISK and EKSPONG (1962), but the experiment of these authors is difficult to interpret.

B. Isospin

ALSTON et al. measured the following branching ratio of the decays of the Λ^*

$$R = \frac{\Gamma(\Lambda^* \to \Sigma^0 \pi^0)}{\Gamma(\Lambda^* \to \Sigma^+ \pi^-) + \Gamma(\Lambda^* \to \Sigma^- \pi^+)} = 0.6 \pm 0.2 . \tag{23.2}$$

The predicted ratio is

$$\begin{aligned}
R &= 2 \quad \text{if} \quad I = 2 \\
R &= 0 \quad \text{if} \quad I = 1 \\
R &= {}^1/_2 \quad \text{if} \quad I = 0 .
\end{aligned} \tag{23.3}$$

The experiment strongly favors $I = 0$. Additional evidence for $I = 0$ is the fact that peaks at 1405 MeV have not been observed in the $\Sigma^+ \pi^0$ and $\Sigma^- \pi^0$ mass spectra. Consistent with $I = 0$ is the apparent absence of the decay $\Lambda^* \to \Lambda + \pi^0$.

C. Spin and parity

Thus far, the Λ^* has always been observed to decay isotropically within experimental error. This fact favors $J = {}^1/_2$ for the Λ^*, but of course does not prove that this assignment is correct. Nothing is known experimentally about the parity. Because of the Minami ambiguity, to obtain information about the Λ^* parity, it must be produced polarized and the polarization of the Λ from its decay must be measured.

DALITZ and TUAN (1960) have suggested that there might exist a bound s state of $K^- p$. If the Λ (1405) is this state, its spin and parity are $J^P = {}^1/_2{}^-$.

24. The Λ (1520) hyperon and the parity of the Σ

A. Mass, width, and isospin of the Λ (1520)

Whereas most excited hyperon states were seen in Dalitz plots of three-body final states, the Λ (1520) was observed as a peak in the total $K^- p$ cross-section measured as a function of energy. FERRO-LUZZI et al. (1962a) obtained the first clear evidence for the existence of the Λ (1520) in looking at $K^- p$ interactions at K^- momenta from ≈ 300 to ≈ 500 MeV/c. They saw peaks in the following modes

$$\begin{aligned}
K^- + p &\to \overline{K} + N \\
&\to \Sigma + \pi \\
&\to \Lambda + 2\pi .
\end{aligned} \tag{24.1}$$

The peaks in the $\Lambda\pi^+\pi^-$ and $\overline{K}^0 n$ modes are shown in Fig. 24.1. The incident K momentum at the peak was 395 MeV/c, corresponding to a mass of 1520 MeV. The branching ratios in the peak were found to be

$$\overline{K}N : \Sigma\pi : \Lambda\pi\pi = 30:55:15 . \qquad (24.2)$$

More complete data on cross sections and a detailed analysis are given in a paper by WATSON, FERRO-LUZZI, and TRIPP (1963). The values of the mass and width from this paper are

$$M = 1519.4 \pm 2 \text{ MeV}, \qquad \Gamma = 16.4 \pm 2 \text{ MeV} .$$

Although these values are quite accurate for a particular experiment, the values are likely to be different in other experiments because of interference effects.

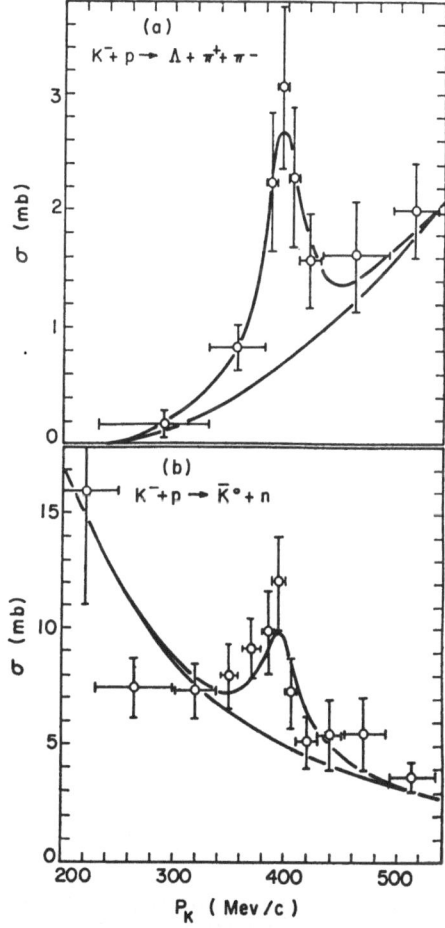

Fig. 24.1. Momentum dependence of the cross section for the reactions (a) $K^- + p \rightarrow \Lambda + \pi^+ + \pi^-$ and (b) $K^- + p \rightarrow K^0 + n$. The lower curves in (a) and (b) represent the presumed non-resonant backgrounds, while the upper curves contain the resonance in addition (FERRO-LUZZI 1962a)

The Λ (1520) has also been seen in three-body final states. BALTAY et al. (1963) have looked at Y* and $\overline{Y}$* production in proton-antiproton collisions. Fig. 24.2 shows a Dalitz plot of $\overline{\Sigma}^{\pm}\pi\Lambda$ and $\Sigma^{\pm}\pi\overline{\Lambda}$ events with projections on the $\Lambda\pi$ (or $\overline{\Lambda}\pi$) and $\Sigma\pi$ (or $\overline{\Sigma}\pi$) axes. Clear evidence occurs for the Λ (1520), the Λ (1405) and the Σ (1385). The antiparticles were also seen in this experiment.

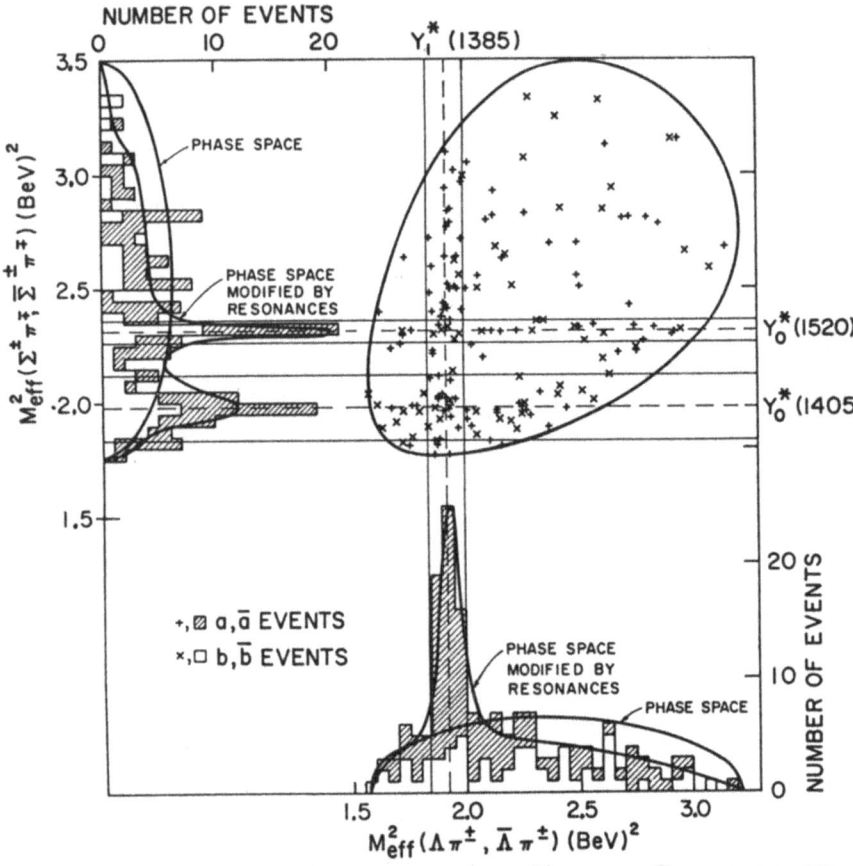

Fig. 24.2. Dalitz plot for squares of effective masses of $\Sigma^{\pm}\pi$ and $\overline{\Sigma}^{\pm}\pi$ vs $\Lambda\pi$ and $\overline{\Lambda}\pi$. Bands 2Γ in width have been drawn through the plot, from 1355 to 1415 MeV for the Σ (1385) or Y*, 1355 to 1455 MeV for the Λ (1405) or Y* (1405) and 1505 to 1535 MeV for the Λ (1520) or Y* (1520). Best-fitting phase-space distributions modified by Y* resonances are drawn in addition to simple phase-space distributions. The modified phase-space curves are computed with the usually accepted positions and widths of the Y* resonances, but with a mass of 1389 MeV and a Γ of 26 MeV for the Σ* (BALTAY 1963)

The Λ (1520) has also been seen in other experiments. Fig. 24.3 from ALEXANDER et al. (1962) shows Dalitz plots of the various charge states of $\Sigma\pi K$ and projections on the $M_{\Sigma\pi}^2$ axes. Clear evidence for the Λ (1520) and the Λ (1405) appears in Fig. 24.3(a). Figures 24.3(b) and (c) show that no evidence appears for charged counterparts of either the Λ (1405) or Λ (1520). This is evidence that both these states have isospin $I = 0$.

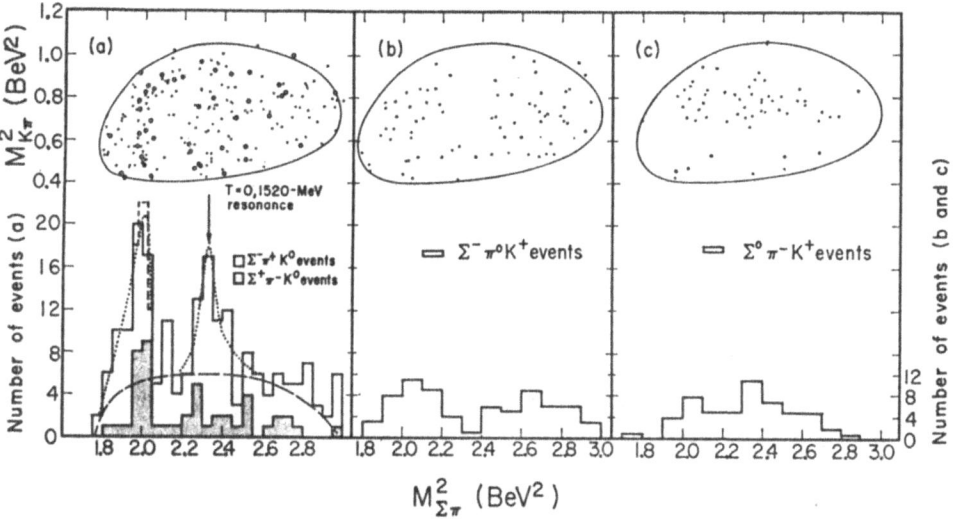

Fig. 24.3. Dalitz plots and distributions in $M^2(\Sigma\pi)$ for $\Sigma\pi K$ events, showing evidence for the Λ (1405) and Σ (1520). In (a) the dashed curve represents the background, and the dotted curves are resonance-type fits. The fact that peaks do not appear in (b) and (c) is evidence that the resonances have isospin zero (ALEXANDER 1962b)

Further evidence that the Λ (1520) is an isosinglet is the absence of the decay Λ (1520) $\rightarrow \Lambda + \pi$, since the $\pi\Lambda$ system is $I = 1$. This is not proof that the Λ (1520) has $I = 0$, however, since the $\Lambda\pi$ decay mode might be inhibited for other reasons. Proof comes from observation by FERRO-LUZZI et al. (1962a) of a peak in the mode $\Sigma^0\pi^0$. This follows because K^-p is a mixture of $I = 0$ and $I = 1$, and $\Sigma^0\pi^0$ has no $I = 1$ component. With the assignment $I = 0$, the branching ratios into the various charge states can be predicted. These are

$$K^-p : \overline{K}^0n = 1:1$$
$$\Sigma^+\pi^- : \Sigma^0\pi^0 : \Sigma^-\pi^+ = 1:1:1 \qquad (24.3)$$
$$\Lambda\pi^+\pi^- : \Lambda\pi^0\pi^0 = 2:1$$

for events above background. The corresponding ratios for background events can be found by measuring the ratios above and below the peak. If one does not obtain these ratios in an experiment it means either that the background ratios are varying through the resonance or that there is interference with background amplitudes. (We are assuming isospin conservation both in production of the peak and in its decay.) The data on the branching ratio $K^-p/\overline{K}^0n$, on angular distributions, and on the polarization of the Σ^+ all indicate that interference with background is present.

B. Spin and parity of the Λ (1520)

The spin of the Λ^* can be measured by looking at the angular distribution in the neighborhood of the peak in the cross section. Interference with background makes this angular distribution a non-unique function of the spin, and a detailed analysis must be performed. Following is a brief summary of the analysis (FERRO-LUZZI 1962a, WATSON 1963).

The observed angular distribution of K^-p and $\overline{K}^0n$ can be written in the form

$$I(\Theta) = a + b\cos\Theta + c\cos^2\Theta . \tag{24.4}$$

Since no powers higher than $\cos^2\Theta$ are required, the spin is most likely $^3/_2$. Spin $^5/_2$ cannot be strictly excluded, even though it seems unlikely. The isotropic term a is large, while the coefficient b of $\cos\Theta$ is small. The simplest explanation of this is that the interference arises chiefly from states of the same parity. At low energy there is a large s-wave phase shift for $\overline{K}$-nucleon scattering, so it is reasonable to assume that the isotropic term is chiefly s wave. If so, then the resonance is a $d_{s/_2}$ state of K^-p, and the parity is negative. With this interpretation, the $b\cos\Theta$ term arises from a small amount of p wave.

C. Parity of the Σ

We assume for this discussion that the parity of the K has been established to be negative. Without this assumption, the statement, "The Σ parity is positive" should be replaced by the statement, "The $K - \Sigma$ relative parity is negative."

Examination of the channel $K^- + p \rightarrow \Sigma + \pi$ in the region of the Λ (1520) has provided evidence that the Σ parity is positive (TRIPP 1962, WATSON 1963). A simplified version of the argument goes as follows: The K^-p resonant chanel is $d_{s/_2}$ and therefore the corresponding state in the $\Sigma\pi$ channel must be either $d_{s/_2}$ or $p_{s/_2}$, depending on the parity. Because of the Minami ambiguity, the angular distribution is independent of which of these possibilities holds. Since both the π and K are pseudoscalar, $d_{s/_2}$ means positive Σ parity. Thus we have the following possibilities

$$d_{s/_2} \rightarrow d_{s/_2},\ s_{1/_2} \rightarrow s_{1/_2}\ \ \text{if}\ \ P_\Sigma = + ,$$

$$d_{s/_2} \rightarrow p_{s/_2},\ s_{1/_2} \rightarrow p_{1/_2}\ \ \text{if}\ \ P_\Sigma = - .$$

Because of the interference between $J = ^1/_2$ and $J = ^3/_2$ amplitudes, the Σ will be produced polarized. The sign of the polarization depends on whether the interference is $(p_{1/_2}, p_{s/_2})$ or $(s_{1/_2}, d_{s/_2})$. Since the decay of the Σ^+ is a highly efficient analyzer, the polarization can be measured.

This does not completely remove the ambiguity, however, since replacing the amplitudes by their complex conjugates also reverses the sign of the polarization. To remove this last ambiguity, TRIPP et al. (1962) invoked the Wigner theorem (1955) which says that as the energy is increased, a resonant amplitude rotates counterclockwise in the complex plane. Although WIGNER proved this theorem for elastic scattering, it should hold in production reactions as well. WIGNER's proof depends on the assumption that the interaction vanishes outside of a certain radius. Although the radius of the interaction is not a Lorentz invariant quantity, WIGNER's assumption does not appear to be a serious shortcoming. However, we do not know of any proof using field theory of a theorem analogous to the Wigner theorem. Assuming the Wigner theorem to hold, the ambiguity is removed and the experiment favors even Σ parity.

The analysis leading to this result is based on the simplest assumptions consistent with the data. Although it is possible that a set of more complicated amplitudes might be consistent with negative Σ parity, no one has produced such a set of amplitudes which fit the data as well as the amplitudes of TRIPP et al. (1962). In a computer search, WATSON et al. (1963) failed to find a fit with odd Σ parity.

25. The Σ (1660) hyperon

A. Mass, width, and isospin

Evidence for the existence of an excited hyperon of mass 1680 MeV was first obtained by BASTIEN et al. (1962b) and by ALEXANDER et al. (1962a). BASTIEN et al. observed it as a peak in the $\Sigma\pi$ channel in K^-p interactions at incident momenta from 293 to 850 MeV/c. It was seen by ALEXANDER et al. in π^-p interactions at 1.89 and 2.04 GeV/c. Since it appeared in the $\Lambda\pi$ as well as the $\Sigma\pi$ channel, it must have $I = 1$.

Confirmatory evidence has come from experiments by ALVAREZ et al. (1963) and by BASTIEN and BERGE (1963). ALVAREZ et al. looked at the reactions

$$
\begin{aligned}
K^- + p &\to \Lambda + \pi^+ + \pi^- \\
&\to \Lambda + \pi^+ + \pi^- + \pi^0 \\
&\to \Sigma^0 + \pi^+ + \pi^- \\
&\to \Sigma^\pm + \pi^\mp + \pi^0 \\
&\to \Sigma^\pm + \pi^\mp + \pi^+ + \pi^- \\
&\to \overline{K}^0 + \pi^- + p \, .
\end{aligned}
\tag{25.1}
$$

These authors observed peaks in the mass distribution in the channels in which a Λ or a Σ were produced, but not in the $\overline{K}N$ channel. Histograms showing the mass distributions of $\Lambda\pi^+$, $\Sigma^+\pi^0$, $\Sigma^0\pi^+$ and $\overline{K}^0 p$ are given in Fig. 25.1.

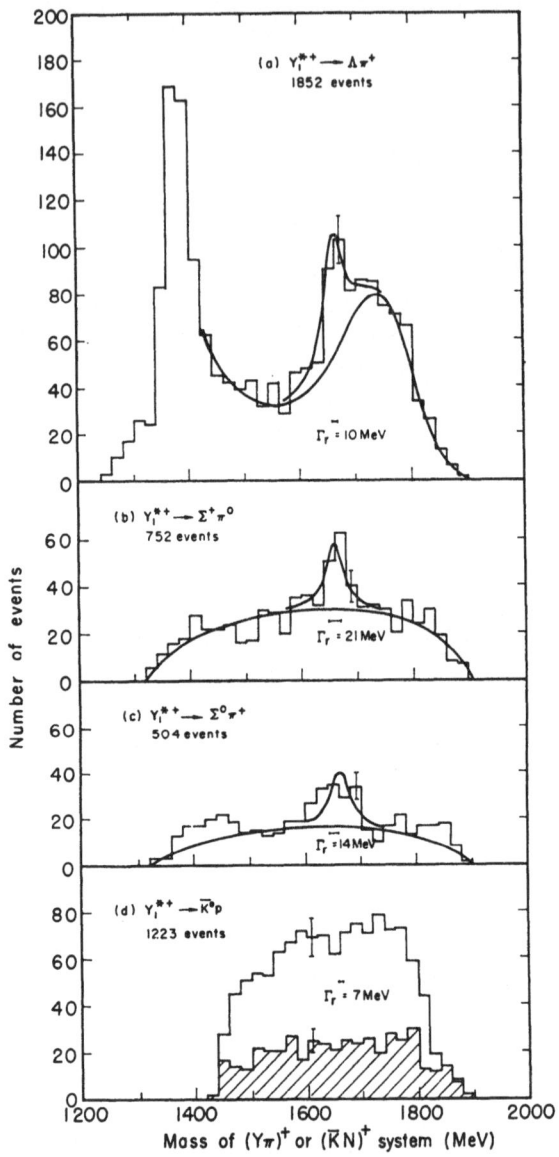

Fig. 25.1. Two-body decay modes of the Σ^+ (1660). The dashed portion of Fig. 25.1 (d) includes only non-K* events (events satisfying the criterion 850 MeV $< M$ (Kπ) < 940 MeV have been removed). The curves represent the best fits of the data to a smooth background curve plus a Breit-Wigner distribution. Here Γ_r indicates the experimental resolution in the 1660 MeV region (ALVAREZ 1963)

The mass spectrum of $\Lambda\pi^+$ is particularly interesting, as can be seen from Fig. 25.1(a). The major part of the peak at a mass of about 1700 is a manifestation of the Σ^- (1385). See the Dalitz plot of Fig. 22.1

to see how this peak arises. However, the Σ (1660) appears superimposed on top of this peak. Fig. 25.1 (b) and (c) show the appearance of the Σ (1660) in the $\Sigma\pi$ mass spectra. Here, there is little or no interference from the Σ (1385) because this latter hyperon rarely decays into $\Sigma + \pi$.

Peaks were also observed in the $\Lambda\pi^+\pi^0$, $\Sigma^-\pi^+\pi^+$ and $\Sigma^+\pi^+\pi^-$ mass distributions. The position and width of the peak varied somewhat from channel to channel. The best estimate for the position and width are

$$M = 1660 \pm 10 \text{ MeV}, \qquad \Gamma = 40 \pm 10 \text{ MeV}.$$

The observed production cross section was

$$\sigma(K^- + p \rightarrow \Sigma^{*+} + \pi^-) = 0.4 \text{ mb} \tag{25.2}$$

with branching ratios

$$\Sigma^{*+} : \Sigma^{*0} : \Sigma^{*-} = 5 : < 1 : < 1 .$$

B. Spin

BASTIEN and BERGE (1963) looked at K^-p interactions at incident K^- momenta of 620, 760, and 850 MeV/c. These momenta correspond to c.m. total energies of 1616, 1681, and 1723 MeV; i.e., at energies below and above the Σ (1660) state. Differential cross sections for various channels were measured and fitted to a power series in $\cos\Theta$. The angular distributions are essentially isotropic below the resonant energy but contain a large $\cos^2\Theta$ term above the resonant energy. The simplest explanation of the $\cos^2\Theta$ term is that it is an interference effect between the resonance and a $J = 1/2$ background. This implies that the spin of the Σ (1660) is greater than $1/2$. The fact that an appreciable $\cos^3\Theta$ term did not appear at a total energy of 1681 MeV makes the most probable assignment $J = 3/2$. However, this value is by no means established.

C. Parity

TAHER-ZADEH et al. (1963) obtained information concerning the parity of the Σ (1660) by looking at the reaction

$$K^- + d \rightarrow p + \pi^- + \Lambda \tag{25.3}$$

with incident momenta of 600, 765 and 850 MeV/c. The authors considered only those events in which the momentum of the emitted proton was less than 285 MeV/c in the laboratory. They assumed that for these events the impulse approximation (CHEW 1952) is valid. In this approximation the proton is treated essentially as a spectator.

As a check of this assumption, the momentum spectrum and angular distribution of the proton were obtained. The momentum spectrum agreed with the expected momentum distribution for protons in the deuteron, assuming that this is given by a Hulthen wave function (HULTHÉN 1951). Also, the angular distribution of the proton was observed to be isotropic in the laboratory, as it should be if the proton is a spectator. Because of these checks, TAHER-ZADEH et al. assumed they were measuring the cross section for the reaction

$$K^- + n \to \Lambda + \pi^- . \qquad (25.4)$$

In this reaction, both the initial and final state have isospin $I = 1$. On the other hand, in the reaction

$$K^- + p \to \Lambda + \pi^0 , \qquad (25.5)$$

the initial state is a mixture of one-half $I = 1$ and one-half $I = 0$. Therefore the cross section for reaction (25.4) should be twice that for reaction (25.5). In fact, the agreement is excellent within the experimental error (TAHER-ZADEH 1963, BASTIEN 1963).

We briefly summarize the evidence concerning the parity of the Σ (1660). First TAHER-ZADEH et al. assumed that the reaction $K^- + n \to \to \Lambda + \pi^-$ could be described in terms of $s_{1/2}$ and $p_{1/2}$ amplitudes, plus the resonant amplitude which was assumed to be either $p_{3/2}$ or $d_{3/2}$. The analysis resembles that of TRIPP et al. (1962) in obtaining the Σ parity. There are two ambiguities. The cross section is invariant, but the polarization changes sign if (1) the parities of all states are changed (Minami ambiguity), or if (2) all amplitudes are replaced by their complex conjugates. The first ambiguity is removed by measurement of the Λ polarization from its decay asymmetry, and the second by invoking the Wigner theorem that the amplitude must rotate counterclockwise in the complex plane as the energy is increased.

TAHER-ZADEH et al. obtained the result that the parity of the $\Sigma(1660)$ is positive; i.e., that the state is a $p_{3/2}$ resonance of a Λ and π. As the authors point out, this conclusion is tentative, as it depends on the impulse approximation, on the assumption that the resonance is $J = {}^3/_2$, and on the assumption that the only other states present are $s_{1/2}$ and $p_{1/2}$.

26. The Λ (1815) and the K^--nucleon cross section

A. The Λ (1815) hyperon

Preliminary evidence for a resonance in the $K^- p$ system at about 1800 MeV came from total cross section measurements by COOK et al. (1961) and by BASTIEN et al. (1961). This resonance was confirmed by more detailed measurements by CHAMBERLAIN et al. (1962). The total $K^- p$ cross section in the neighborhood of the resonance is shown in Fig. 26.1 (a). The resonance has a mass and width given by

$$M = 1815 \text{ MeV}, \qquad \Gamma = 120 \text{ MeV} .$$

As with other broad peaks the mass and especially the width are uncertain because of interference of the resonance amplitude with background.

In Fig. 26.1(b) is shown the K⁻n total cross section σ_n obtained from the K⁻d and K⁻p total cross sections σ_d and σ_p by using an im-

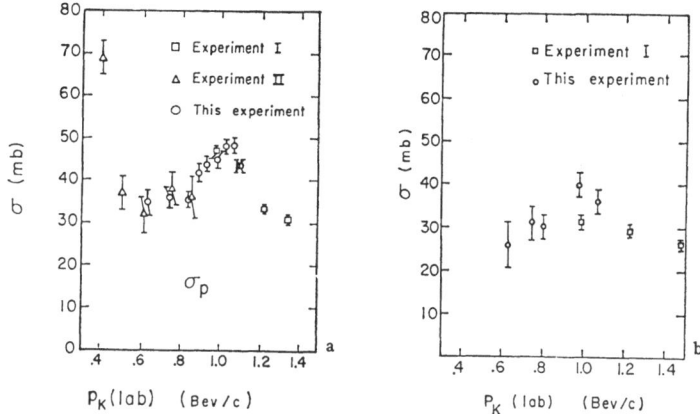

Fig. 26.1. (a) The total K⁻p cross section, showing evidence for the Λ (1815). (b) The total K⁻n cross section. "Experiment I" refers to Cook et al. (1961), "Experiment II" to Bastien et al. (1961) and "This Experiment" to Chamberlain et al. (1962)

pulse approximation with screening. The formula used (Glauber 1955) is

$$\sigma_n = \frac{\sigma_d - \sigma_p}{1 - (\sigma_p/4\pi)\langle 1/r_d^2\rangle}, \qquad (26.1)$$

where $\langle 1/r_d^2 \rangle$ is the inverse mean square radius of the deuteron. The screening affected σ_n by about 10%. There is some evidence for a slight bump in σ_n at around the same energy as the K⁻p peak. However, the magnitude of the rise is much below that required by charge independence if the peak is attributed to an $I = 1$ resonance. Theoretically

$$\sigma_n/\sigma_p = 2 \quad \text{if} \quad I = 1, \quad \sigma_n/\sigma_p = 0 \quad \text{if} \quad I = 0. \qquad (26.2)$$

Since experimentally, $\sigma_p > \sigma_n$, the data favor $I = 0$.

Information about the spin of the resonance comes from elastic scattering (Beall 1962) and charge exchange (Ferro-Luzzi 1962) measurements. Terms of $\cos^5\Theta$ were necessary to fit the elastic scattering angular distribution, and terms of $\cos^6\Theta$ were necessary in the charge exchange reaction at an energy somewhat above the resonance. Since there is a peak in the $\cos^5\Theta$ term near the resonant energy, it is assumed that this term arises from interference between the resonant amplitude and a non-resonant term. This makes the most probable spin of the resonance to be $J = 5/2$. The unitarity limit on the height of the peak is consistent with this assignment. However, other values of the spin, especially $J = 7/2$, are not excluded by the data. The parity is presently unknown.

B. An effect at mass 1765 MeV

BARBARO-GALTIERI et al. (1963) have studied the reaction

$$K^- + n \to K^- + p + \pi^-$$

by looking at K^- interactions in deuterium and using the impulse approximation. In a mass-squared plot of the K^-p distribution, these authors observed, in addition to the Λ (1520), a peak at 1765 MeV. It is not ruled out that this 1765 MeV peak is the Λ (1815) shifted by a statistical fluctuation or by other effects. However, the authors suggest that a $\overline{K}N$ resonance at this mass is also possible.

If so, the broad peak at 1815 MeV is really two interfering resonances. A tentative assignment of quantum numbers $I = 1$, $J = {}^5/_2$ is made. The spin assignment comes from an interpretation of the $\cos^5\Theta$ term in the K^-p angular distribution observed by CHAMBERLAIN et al. (1962). This term could arise from $d_{{}^5/_2} - f_{{}^5/_2}$ interference, as the authors suggest, or from other causes. If the Λ (1815) is $f_{{}^5/_2}$ or $J^P = {}^5/_2{}^+$, as has been suggested on theoretical grounds, the effect at 1765 has negative parity. There is no experimental evidence on this point.

The isospin assignment of the peak at 1765 MeV comes from an interpretation of K^-p elastic scattering and charge-exchange data. But the presently available data do not require this interpretation.

We also mention that BELYAKOV et al. (1962) obtained weak evidence for a hyperon of mass $M = 1760$ MeV in π^-p interactions with a 7 GeV/c π^- beam in propane. The decay mode observed was

$$Y^* (1760) \to \Sigma (1385) + \pi^-, \qquad \Sigma^* \to \Lambda + \pi^+. \tag{26.3}$$

C. High energy K^-p cross section

The K^-p total cross section has been measured at higher energies. A summary of the data are given in Fig. 26.2 (DIDDENS 1963a). There is

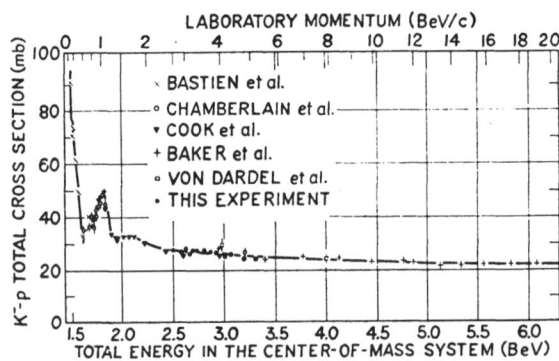

Fig. 26.2. The K^-p total cross section as a function of the total energy in the c.m. system. The references in the figure can be found from the paper of DIDDENS et al. (1963a)

evidence for a slight hump or "shoulder" at 2200 MeV, but its significance is unknown.

27. The Ξ (1530) hyperon

A. Mass and width

The Ξ (1530) or Ξ* was seen by PJERROU et al. (1962) and by BERTANZA et al. (1962a). We consider first the experiment by PJERROU et al., in which 1.8 GeV/c incident K⁻ mesons produced Ξ particles in the reactions

$$K^- + p \rightarrow \Xi^- + \pi^0 + K^+$$
$$\rightarrow \Xi^- + \pi^+ + K^0 . \tag{27.1}$$

Most of the 88 observed events of this type corresponded to

$$K^- + p \rightarrow \Xi^{*-} + K^+ \tag{27.2}$$

and

$$K^- + p \rightarrow \Xi^{*0} + K^0. \tag{27.3}$$

Dalitz plots of the events are shown in Fig. 27.1 and an effective mass

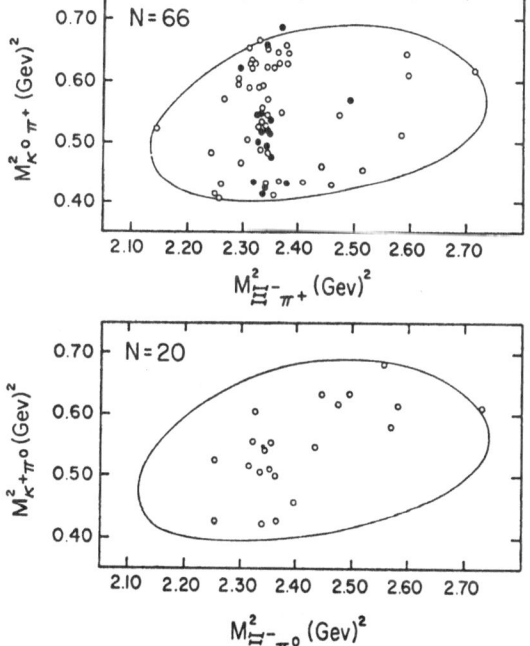

Fig. 27.1. Dalitz plots for the reactions $K^- + p \rightarrow \Xi^- + \pi^+ + K^0$ and $K^- + p \rightarrow \Xi^- + \pi^0 + K^+$. Solid dots represent events with both charged Λ and K⁰ decays (PJERROU 1962)

plot in Fig. 27.2. The mass and width are given by

$$M = 1530 \text{ MeV}, \qquad \Gamma = 7 \pm 2 \text{ MeV} .$$

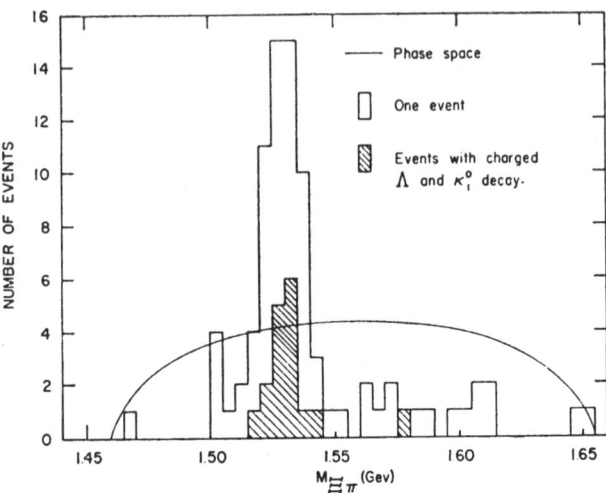

Fig. 27.2. Effective mass plot for $\Xi\pi$ systems observed in K^-p reactions. The line is the relativistic three-body phase space (PJERROU 1962)

B. Isospin

Since background events were infrequent in Ξ^* production, the branching ratio

$$R_1 = (\Xi^- \pi^0 K^+)/(\Xi^- \pi^+ K^0) \qquad (27.4)$$

gives information about the isospin. If the Ξ^* has $I = {}^3/_2$, it can be reached only from the $I = 1$ component of K^-p. Then R_1 is given by the unique value $R_1 = 2$. The ratio R_1 cannot be predicted for $I = {}^1/_2$, since then the Ξ^* can be produced in both the $I = 0$ and $I = 1$ states of K^-p. Experimentally $R_1 = {}^{30}/_{75}$, showing $I = {}^1/_2$. Both the production of the Ξ^* and its decay were isotropic in this experiment within experimental error.

BERTANZA et al. saw the Ξ^* in the same interaction at incident K^- momenta of 2.24 and 2.5 GeV/c. They found a mass $M = 1535 \text{ MeV}$ and width $\Gamma < 35 \text{ MeV}$. These authors obtained for the ratio $R_1 = {}^2/_{10}$ again favoring $I = {}^1/_2$. Further evidence concerning the isospin comes from observation of the branching ratios

$$R_2 = \frac{\Gamma(\Xi^{*0} \to \Xi^- \pi^+)}{\Gamma(\Xi^{*0} \to \Xi^0 \pi^0)}, \qquad R_3 = \frac{\Gamma(\Xi^{*-} \to \Xi^0 \pi^-)}{\Gamma(\Xi^{*-} \to \Xi^- \pi^0)} . \qquad (27.5)$$

The expected ratios are

$$R_2 = R_3 = 2 \quad \text{if} \quad I = {}^1/_2, \qquad R_2 = R_3 = {}^1/_2 \quad \text{if} \quad I = {}^3/_2 . \qquad (27.6)$$

Experimentally $R_2 = {}^5/_0$, $R_3 = {}^3/_2$, favoring $I = {}^1/_2$. Background events were included in these experimental ratios.

C. Spin and parity

SCHLEIN et al. (1963) measured the decay angular distribution $I(\theta)$ and polarization $P(\theta)$ of Ξ^*'s decaying into the mode

$$\Xi^{*0} \to \Xi^- + \pi^+.$$

They used the method of analysis of BYERS and FENSTER (1963), which, in principle, enables one to determine the spin and parity. In practice, the method did not yield a unique answer because of insufficient events. The first conclusion reached by the authors is that the spin of the Ξ^* is $J \geqq {}^3/_2$. The second conclusion is that the data are compatible with the Ξ^* being a resonant $\Xi\pi$ state with $J^P = {}^3/_2{}^+$ ($p_{3/_2}$) but that $d_{3/_2}$ is unlikely (with probability 0.016). Thus, if $J = {}^3/_2$, then the relative $\Xi^*\Xi$ parity is positive. However $J > {}^3/_2$ cannot be excluded by the data.

Other data; e.g., the data of CONNOLLY et al. (1963), have supported the conclusion that $J \geqq {}^3/_2$. Furthermore, no evidence exists that would lead one to believe that the spin were greater than $^3/_2$. Then, assuming that the parity of the Ξ is positive, the most probable assignment for the Ξ^* is $J^P = {}^3/_2{}^+$.

27 a. The Ω hyperon

A large group of physicists at Brookhaven National Laboratory (BARNES 1964) have observed a hyperon of hypercharge $Y = -2$ which decays by weak interactions. The existence of this hyperon was predicted (BEHRENDS 1962, GELL-MANN 1962b, GLASHOW 1962) within the framework of unitary symmetry (to be discussed in the next section). The other quantum numbers of the Ω are predicted to be $I = 0$, $J^P = {}^3/_2{}^+$.

The first Ω seen was observed in the following reaction

$$K^- + p \to \Omega^- + K^+ + K^0$$

with 5.0 GeV/c K$^-$ mesons incident in an 80 inch hydrogen bubble chamber. The measured mass of the Ω is

$$M = 1685 \pm 12 \text{ MeV}$$

and its lifetime (based on one event) is

$$\tau \approx 0.7 \times 10^{-10} \text{ sec}.$$

Two observed decay modes of the Ω are

$$\Omega^- \to \Xi^0 + \pi^-$$
$$\to \Lambda + K^-$$

Other decay modes should also exist.

28. A classification scheme

Many of the observed mesons and baryons can be classified within the framework of representations of the unitary group SU_3. Still more particles can be accommodated if the idea is adopted that particles lie on Regge trajectories. Good discussions of Lie groups, including SU_3, have been given by BEHRENDS et al. (1962), HAIG et al. (1963), GOURDIN (1964), and others. Regge poles and trajectories have been discussed at length by FRAUTSCHI (1964), OMNÈS and FROISSART (1963) and others. We shall not explain the underlying ideas behind either unitary symmetry or Regge trajectories, but shall just state the rules of classification.

The representation of SU_3 which is most successful is the eight dimensional representation, or the "eightfold way" (GELL-MANN 1961, 1962, NE'EMAN 1961). A representation which was proposed earlier but which at present seems less successful is the original SAKATA (1956) model or threefold way. If the Λ and Σ should turn out to have opposite parity, or if the Ξ should turn out to have quantum numbers other than $J^P = \frac{1}{2}^+$, the eightfold way will have to be abandoned (with respect to $N\Lambda\Sigma\Xi$, at least) in favor of some other way.

The numerology of the eightfold way is that from the product of two 8-representations one can form representations of dimension 1, 8, 10, and 27 as follows

$$8 \times 8 = 1 + 8 + 8 + 10 + \overline{10} + 27 . \qquad (28.1)$$

The two 8's have different symmetry properties, and the members of the 10 and $\overline{10}$ have opposite values of the hypercharge. The isospin and hypercharge of the various members of the unitary multiplets are given in Table 28.1.

Table 28.1. Isospin and hypercharge of SU_3 supermultiplets formed from 8×8

Dimension of representation	Isospin and hypercharge of submultiplets[1]								
1	I 0								
	Y 0								
8	I $1/2$	0	1	$1/2$					
	Y 1	0	0	-1					
10	I $3/2$	1	$1/2$	0					
	Y 1	0	-1	-2					
10	I $3/2$	1	$1/2$	0					
	Y -1	0	1	2					
27	I 1	$1/2$	$3/2$	0	1	2	$1/2$	$3/2$	1
	Y 2	1	1	0	0	0	-1	-1	-2

[1] The hypercharge of an isospin multiplet is given directly below the isospin. The isospin multiplets are representations of SU_2.

If this scheme has anything to do with reality, then we should expect the mesons and baryons to occur in multiplets of 1, 8, 10, and 27, (singlets, octuplets*, decuplets and 27-plets) where all members of a given multiplet have the same spin and parity. (The original Sakata model

---

* The usual notation in the literature for a multiplet of eight particles is "octet." See footnote 3 of the paper by GLASHOW and ROSENFELD (1963) for justification of the notation "octuplet."

gives $3 \times \bar{3} = 1 + 8$.) To the extent that unitary symmetry holds, it is more than just a classification scheme. For example, one can predict branching ratios for the production and decays of the members of a given supermultiplet, just as isospin symmetry enables one to predict branching ratios within an isospin multiplet.

If unitary symmetry were exact, all members of a multiplet would have the same mass. This state of affairs is clearly not realized in nature. However, under a certain assumption about the way unitary symmetry is broken, a perturbation theory formula can be derived (OKUBO 1962, GELL-MANN 1962b) which relates the masses of the members of a given multiplet

$$M = M_0(1 + a\,Y + b\,[I(I + 1) - \tfrac{1}{4}\,Y^2]) \qquad (28.2)$$

where M_0, a, and b are constants, I is the isospin, and Y is the hypercharge (which is related to the strangeness and baryon number by $Y = S + B$). The above formula is supposed to hold for baryons. A similar formula is supposed to hold for mesons, but it has been suggested that M and M_0 should be replaced by M^2 and M_0^2 for mesons. GLASHOW and SAKURAI (1962) have suggested that this formula also applies to the widths of unstable particles, suitably modified by phase space factors.

The ground state baryons ($N\Lambda\Sigma\Xi$) seem to fit nicely into an octuplet with $J^P = \tfrac{1}{2}^+$. (To get the multiplicity, all charge states are counted: $pn\Lambda\Sigma^+\Sigma^0\Sigma^-\Xi^0\Xi^-$.) The mass formula simplifies for an octuplet. For the ground state baryons it becomes

$$\frac{M_N + M_\Xi}{2} = \frac{3\,M_\Lambda + \Sigma\,M}{4}. \qquad (28.3)$$

This formula agrees with the experimental masses of the baryons to within 1%. The pseudoscalar mesons ($\pi K \eta$) also can be classified as an octuplet, fitting the mass formula to 3%.

For a decuplet, the mass formula is particularly simple, giving an equal spacing rule. The masses of the Δ (1238), Σ (1385), Ξ (1530) and Ω (1685) fit this rule remarkably well. Also, although the spins and parities of the Σ (1385) and Ξ (1530) are not definitely established, their most probable assignments are $J^P = \tfrac{3}{2}^+$, just like the Δ. The quantum numbers of the Ω are not yet known.

The mass formula holds to 3% or better for the baryon and meson octuplets and for the baryon decuplet. Why this should be so is somewhat mysterious, as the mass splittings are so large that lowest order perturbation theory is not expected to be a good approximation.

The classification of the nine known vector mesons ($\rho\omega\,\varphi K^*$) presents a problem. It has been usual to classify these nine particles as being an octuplet and a singlet. However, the ω and φ have the same quantum numbers, so it is not clear which should be a singlet and which a member of the octuplet. We consider four possibilities, not meant to be exhaustive.

1. If one considers A parity to be a good quantum number, then the ω is the singlet, as it has been assigned $A = -$ while the ρ, φ, and K^* were assigned $A = +$.

2. The eight vector mesons do not fit the octuplet mass formula whether the φ or the ω belongs to the octuplet. If the mass of the $I = 0$ vector meson is calculated from the mass formula, the mass comes out between the masses of the ω and φ. For this reason, a number of authors (SAKURAI 1952, KATZ 1963) have discussed the possibility that two other states, the φ_0 and ω_0 belong to the octuplet and the singlet representations, and the φ and ω are mixtures of these states.

3. It has been suggested that other vector mesons exist, so that the vector mesons belong to a higher dimensional representation of SU_3 such as the 27. Alternatively (GLASHOW 1963a, MINAMI 1963), there may exist two octuplets of vector mesons.

4. The vector mesons may be representations of another group. In particular, TARJANNE and TEPLITZ (1963) have suggested that they are representations of SU_4.

The unitary symmetry scheme can be combined with the idea of Regge poles and trajectories (REGGE 1959, 1960, CHEW 1961, 1962). If spins of the different particles are plotted versus their masses, the points representing some of the particles can be connected by smooth lines or Regge trajectories. All particles on a given trajectory are supposed to have the same quantum numbers other than spin, and the spins are supposed to increase in steps of two. Unfortunately, little is known about the slope of a Regge trajectory or about how many particles should lie on a given trajectory. Therefore, at present, the classification scheme amounts to little more than guesswork.

In Tables 28.2 through 28.5 we give a tentative classification of some of the baryons and mesons according to the SU_3-Regge scheme.

Table 28.2. A baryon octuplet with $J^P = 1/2^+$ and possible Regge recurrences. The multiplicity 8 is arrived at by counting all charge states. A question mark after the quantum numbers of a particle means that the quantum numbers have not been definitely established

Symbol	N	Λ	Σ	Ξ
Mass MeV	939	1115	1193	1320
I, J^P	$1/2, 1/2^+$	$0, 1/2^+$	$0, 1/2^+$	$1/2, 1/2^+$?
Y, S	1, 0	0, —1	0, —1	—1, —2
Old symbol	$N^*_{1/2}$	Y^*_0		
New symbol	N^*	Λ^*		
Mass MeV	1688	1815		
Width MeV	100	120		
I, J^P	$1/2, 5/2^+$?	$0, 5/2^+$?		
Y, S	1, 0	0, —1		

There are still a number of mesons and baryons left over after this classification. The Λ (1405) with $I = Y = 0$ is usually classified as a unitary singlet. Some of the high mass N* and Y* states may be Regge recurrences of states of lower mass. But we have left unclassified such baryon states as N (1512), Σ (1660), and Λ (1520), and such meson states as the f, B, and $\varkappa$. When more particles are discovered and when the quantum numbers of the existing ones are better known, perhaps all these particles will fit nicely into unitary multiplets – and perhaps not!

Table 28.3. A possible baryon decuplet and Regge recurrence

Old symbol	$N^*_{3/2}$	Y^*_1	$\Xi^*_{1/2}$	
New symbol	Δ	Σ^*	Ξ^*	Ω
Mass MeV	1238	1385	1530	1685
Width MeV	90	50	7	0
I, J^P	$3/2, 3/2^+$	$1, 3/2^+$?	$1/2, 3/2^+$?	0?, ?
Y, S	$1, 0$	$0, -1$	$-1, -2$	$-2, -3$

Old symbol	$N^*_{3/2}$			
New symbol	Δ^*			
Mass MeV	1920			
Width MeV	200			
I, J^P	$3/2, 7/2^+$?			
Y, S	$1, 0$			

Table 28.4. An octuplet of pseudoscalar mesons

Symbol	π	η	K	$\overline{K}$
Mass MeV	138	548	496	496
I^G, J^P	$1^-, 0^-$	$0^+, 0^-$	$1/2, 0^-$	$1/2, 0^-$
$Y = S$	0	0	1	-1

Table 28.5. A possible octuplet of vector mesons and a possible singlet

Symbol	Octuplet				Singlet
	ρ	φ	K^*	$\overline{K}^*$	ω
Mass MeV	750	1019	888	888	784
Width MeV	120	3	50	50	9
I^G, J^P	$1^+, 1^-$	$0^-, 1^-$	$1/2, 1^-$	$1/2, 1^-$	$0^-, 1^-$
$Y = S$	0	0	1	-1	0

In any case, there remains a challenge to experimental physicists to obtain the data which will either confirm this general classification scheme or lay it to rest.

29. Possible violation of CP invariance*

A. Long-lived neutral K decay into two pions

In a precision counter experiment CHRISTENSON et al. (1964) observed evidence that the long-lived neutral K meson decays into two pions. The most straightforward explanation for this result is that CP is not conserved in the decay of the K^0 meson, as pointed out by CHRISTENSON et al. Since the K_1 and K_2 were previously defined as eigenstates of CP, it is useful to define K_S and K_L as the short and long lived components of the K^0. If CP is not conserved, then K_L and K_2 are not identical

---

* Sections 29 through 33 were added in March, 1965.

(and similarly K_S is not the same as K_1). CHRISTENSON et al. observed a branching ratio of the K_L into 2 pions given by

$$R = \frac{K_L \to \pi^+\pi^-}{K_L \to \text{all charged modes}} = (2.0 \pm 0.4) \times 10^{-3}.$$

Some additional evidence for this effect was found by ABASHIAN et al. (1964). Further evidence was found by GALBRAITH et al. (1965) who obtained a branching ratio

$$R = (2.08 \pm 0.35) \times 10^{-3}.$$

In order to "save" CP invariance, a number of other possible explanations for the phenomenon have been proposed. Some of these have been summarized by FEINBERG (1964); we shall also briefly discuss some of the proposals.

B. Possible hypercharge field

It has been proposed that an almost-conserved vector field is coupled to hypercharge, in analogy with the coupling of the electromagnetic field to electric charge (BELL 1964, BERNSTEIN 1964). Another possibility is that the field is a scalar field (LEE 1965). Since the galaxy is made up primarily of nucleons with hypercharge $Y = 1$ (rather than of anti-nucleons with $Y = -1$), the potential energy of the $K^0 (Y = 1)$ would be opposite in sign to that of the $\overline{K}^0$ ($Y = -1$). Such an energy difference between K^0 and $\overline{K}^0$ would lead to regeneration of the K_S even in an apparent vacuum in the laboratory, and thus the apparent K_L decay into two pions would be regenerated K_S decay. Thus CP would not be violated.

A test of this possibility was proposed by the authors, who pointed out that the K_L decay rate into two pions would go as the square of the total energy of the K with respect to the galaxy. This test has already been carried out. In the experiment of CHRISTENSON et al. the K_L beam had an average momentum of 1.1 GeV/c, while in the experiment of GALBRAITH et al. the momenta of the K_L's varied between 1.5 and 5.0 GeV/c. Yet within statistics the decay rates in the two experiments were the same[*].

C. Possible non-exponential decay

According to the Breit-Wigner theory, unstable particles decay according to an exponential law. However, this is an approximate theory

---

[*] The new CERN experiment at 10.7 GeV/c (DE BOUARD et al. 1965) gives for the above ratio $R = (3.5 \pm 1.4) \, 10^{-3}$ (added by the editor).

which is not expected to hold for times extremely long compared to the mean life (see e.g. MATTHEWS 1959). Despite this fact, no one has yet put forth a model of the K_S which would allow sufficiently many K_S mesons to survive to explain the experiments.

D. Does another particle exist?

A number of suggestions have appeared in print that perhaps another particle is responsible for the apparent decay of the K_L into 2 pions. Variations of such suggestions have been given by so many people that we do not give any references.

One possibility is that the K_L decays into a particle X as follows

$$K_L \to X + \gamma, \quad X \to 2\pi \,.$$

The difficulty with this explanation is that the X must have the same mass as the K_L within about an MeV in order to explain the experiments. Also, why is not the X seen in other processes? Another possibility is that the K_L beam is contaminated by another particle X'. But again, the X' must have approximately the same mass and lifetime as the K_L.

E. Are pions bosons?

If an antisymmetric state of two pions could exist, this state would have $CP = -1$, and the K_L could decay into it. Since symmetric two-pion states certainly exist, the discovery of an antisymmetric state would imply that pions obey statistics intermediate between BOSE and FERMI. In connection with this, BLUDMAN (1964) has suggested that perhaps an antisymmetric state of $\pi^+ \pi^-$ exists, but not an antisymmetric state of $\pi^0 \pi^0$. A measurement of the decay rate $K_L \to 2\pi^0$ would be valuable independently of whether BLUDMAN's suggestion is correct. However, the experiment is extremely difficult to carry out.

If pions are not bosons, it is hard to understand why symmetric two-pion states have appeared in a wide variety of interactions, whereas antisymmetric states have appeared very weakly, if at all. Except where otherwise stated, we assume that pions are bosons.

F. Models with CP violation

A large number of different models have been proposed in which CP is violated, either directly in the $K_L \to 2\pi$ channel or in other channels. We shall not take the space to mention all the proposed models.

In a model by SACHS (1964), CP is maximally violated in hypercharge-changing leptonic decays with $\Delta Y = -\Delta Q$. The indirect effects lead to a small $K_L \to 2\pi$ decay rate. The advantage of SACHS' model is that a small violation is replaced by an aesthetically more satisfying maximal violation. The disadvantage is that there is no evidence for decays violating the $\Delta Y = \Delta Q$ selection rule. Some of the experiments on K^0 mesons may have to be reanalyzed without the assumption of CP invariance.

WU and YANG (1964) have discussed CP non-invariance phenomenologically, suggesting additional experiments to explore the characteristics of the CP violating interaction.

In nearly all the discussions of CP violation, it has been assumed that the CPT theorem holds. This assumption implies that T is also violated, and experiments to look for possible violation are under way. No one (to my knowledge) has proposed a specific theory in which CPT does not hold. The difficulty in constructing such a theory may mean that some of the assumptions of quantum mechanics may have to be modified if CPT is found to be violated.

Finally, we mention that the arguments given in previous sections which depend on CP invariance are still essentially correct, since the extent of the violation is so small.

G. Possible short-lived neutral K decay into three pions

Consider the possible decay mode $K_S \to \pi^+ \pi^- \pi^0$. The three-pion final state may have CP either even or odd, depending on the orbital angular momenta of the pions. Let L_1 be the orbital angular momentum of the $\pi^+ \pi^-$ pair and let L_2 be the orbital angular momentum of the π^0 with respect to the c.m. of the pair. Since the K has spin 0, we must have $L_1 = L_2 = L$. The $\pi^+ \pi^-$ state has $CP = +$ independent of L. The π^0 has $CP = -$, and the parity of the orbital motion of the π^0 is $P = (-1)^L$. Therefore, the three-pion state has

$$CP = (-1)^{L+1}.$$

This means that if CP is conserved in the decay of the K_S, the pions must be emitted in states with odd L, most likely in the state with $L = 1$. However, if CP is not conserved, the pions can be emitted in even states, most likely in the state with $L = 0$. In the latter case, and if in addition CP is maximally violated, the decay rate of the K_S will be considerably enhanced because of the absence of angular momentum barriers in the $L = 0$ state. The decay rate for $K_S \to \pi^+ \pi^- \pi^0$ might then be comparable to the decay rate for $K_L \to \pi^+ \pi^- \pi^0$.

A possible CP non-conserving decay $K_S \to \pi^+ \pi^- \pi^0$ was looked for by ANDERSON et al. (1965). Let the ratio R of the amplitude for this decay to the amplitude for the decay $K_L \to \pi^+ \pi^- \pi^0$ be $R = x + iy$. Then the experimental result is

$$x = 0.25 \pm 0.55, \quad y = 0.80 \pm 0.55 .$$

Within statistics, both x and y are consistent with zero. It should be emphasized that even if it should turn out that x and y are of the order of unity, this does not prove that CP is violated. There might be an unknown dynamical factor which enhances the $L = 0$ decay of the K_S. We can distinguish between $L = 0$ and $L = 1$ by looking at a Dalitz plot of the $\pi^+ \pi^- \pi^0$ decays. If $L = 0$, the density of points will be approximately uniform, while if $L = 1$, the center region of the plot will be depopulated. Such an experiment, of course, is not easy to do.

As another test of CP invariance, we point out that the state $3\pi^0$ has $CP = -1$, since for identical spinless bosons L must be even. Therefore if CP is conserved, the decay $K_S \rightarrow 3\pi^0$ will be forbidden. Unfortunately, it would be difficult to detect this mode even if it occured.

30. Recently discovered meson states

A good summary of the evidence concerning newly discovered meson states has been given by GOLDHABER (1965). He goes into considerably more detail than we do, and includes evidence for possible states we do not discuss. He also summarizes evidence on recently discovered baryon states (to be discussed in the next section).

A. The η' meson

In section 19 G, we briefly noted the possible existence of a neutral meson of mass about 960 MeV. Subsequent experiments have confirmed the existence of this state, called the η', with mass and width given by

$$M = 959 \pm 2 \text{ MeV}, \quad \Gamma < 4 \text{ MeV} .$$

The η' has been seen in $K^- p$ interactions (KALBFLEISCH 1964a, 1964b, GOLDBERG 1964a, 1964b, DAUBER 1964). The following decay modes (with branching fractions in parentheses) were observed (KALBFLEISCH 1964b):

$$
\begin{aligned}
\eta' &\rightarrow \eta + \pi^+ + \pi^-, \ \eta \rightarrow \pi^+ + \pi^- + \pi^0 \quad &(0.12 \pm 0.02) \\
&\rightarrow \eta + \pi^+ + \pi^-, \ \eta \rightarrow \text{neutrals} \quad &(0.36 \pm 0.05) \\
&\rightarrow \pi^+ + \pi^- + \gamma \quad &(0.22 \pm 0.04) \\
&\rightarrow \pi^+ + \pi^- + \text{neutrals} \quad &(0.05 \pm 0.04) \\
&\rightarrow \text{neutrals} \quad &(0.25 \pm 0.05)
\end{aligned}
$$

From analysis of decay correlations in the $\pi^+\pi^-\gamma$ mode, KALBFLEISCH et al. (1964b) obtain for the most likely quantum numbers of the η'

$$J^P = 0^-, \quad I^G = 0^+, \quad C = +.$$

These are the same quantum numbers as those of the η. However, the η and η' can be distinguished in the SU_3 scheme, the former belonging to an octuplet representation and the latter to a singlet.

With the quantum numbers $J^P = 0^-$, $I^G = 0^+$, the decay $\eta' \to \eta\pi\pi$ goes via strong interactions. The decay $\eta' \to \pi^+\pi^-$ + neutrals probably consists chiefly of the mode $\eta' \to \eta\pi^0\pi^0$ ($\eta \to \pi^+\pi^-\pi^0$), and the decay $\eta' \to$ neutrals probably is a mixture of $\eta' \to 2\gamma$ and $\eta' \to \eta\pi^0\pi^0$ ($\eta \to$ neutrals). The decay $\eta' \to 3\pi$ is forbidden by conservation of G parity, while the decay $\eta' \to \gamma\pi^0\pi^0$ is forbidden by charge conjugation invariance.

B. The A_1 and A_2 states

We noted in section 19 E a peak in the mass spectrum of ρ and π. This peak has subsequently been resolved into two peaks (CHUNG 1964, ADERHOLZ 1964), called A_1 and A_2, with masses and widths

$$A_1: \ M = 1090 \text{ MeV}, \ \Gamma = 125 \pm 25 \text{ MeV}$$

$$A_2: \ M = 1310 \text{ MeV}, \ \Gamma = \ 80 \qquad \text{MeV}$$

The A_1 peak may be the result of an s-wave $\rho\pi$ interaction, or some other effect rather than a resonant state. However, we shall assume for the discussion that the A_1 is a resonance. For references to the earlier papers see LANDER et al. (1964). Both the A_1 and A_2 have been seen in singly charged states, and therefore have isospin $I \geq 1$. Since they decay into $\rho\pi$, they have $G = -1$.

CHUNG et al. (1964) observed peaking in the K^-K_1 and K_1K_1 mass spectra in the region of 1300 MeV. If this peaking arises from the decay of the A_2 meson, additional information can be obtained about its quantum numbers. A K^-K_1 state has $I = 1$; a K_1K_1 state has $J^P = 0^+$, 2^+, 4^+ etc. But a 0^+ state cannot decay into $\rho + \pi$ without violating parity conservation. Therefore the lowest value of the spin consistent with the data is $J = 2$. Thus, the quantum numbers of the A_2 probably are $I^G = 1^-$, $J^P = 2^+$.⋆

The quantum numbers of the A_1 are at present unknown.

⋆ ALITTI et al. (1965) found that $J^P = 2^+$ is clearly favored for A_2. In the case of A_1 the most probable quantum numbers deduced from the $\rho\pi$ decay mode: $J^P = 1^+$ are in contradiction with the $\eta\pi$ decay analysis (added by the editor).

C. A possible s-wave $\pi\pi$ resonance near 750 MeV

In our discussion of the ρ meson, we pointed out that the forward-backward asymmetry in the decay of the ρ^0 is evidence for a strong s- or d-wave $\pi\pi$ interaction, remarking that an s-wave resonance is not excluded by the data. The possible existence of such a resonance was already noted by HAGOPIAN and SELOVE (1963). Subsequently, other authors interpreted the data in terms of a s-wave $\pi\pi$ state near 750 MeV. In a paper of DURAND and CHIU (1965), the name ε was suggested for this possible state. (See this paper for earlier references.) The quantum numbers of the ε (if it exists) are

$$I^G = 0^+, \quad J^P = 0^+ .$$

D. Status of the ABC, σ and ζ

In section 20, we discussed some inconclusive evidence for a number of $Y = 0$ meson states. Among them are the following.
1. An s-wave $I = 0$ $\pi\pi$ interaction at 310 MeV, called the ABC.
2. An s-wave $I = 0$ $\pi\pi$ resonance at 390 MeV, called the σ.
3. An $I = 1$ $\pi\pi$ resonance at 570 MeV, called the ζ.

Although these states have been subsequently looked for in many experiments, they have not been definitely seen. Therefore, they should be considered as more doubtful than previously thought. We discuss each of them briefly in turn.

Since the ABC has appeared only in an interaction involving He in the final state, it is very possible that the effect is a manifestation of a three-body process involving two pions and He, rather than a property of the $\pi\pi$ system (KLEIN 1965).

Evidence for the existence of the σ meson has been summarized by BROWN and FAIER (1965). Most of this evidence is indirect and therefore is not compelling. No convincing direct evidence for the σ has been observed.

No further evidence in favor of the ζ has appeared. The observed peaking in the $\pi\pi$ system in certain reactions is probably caused by many-body effects rather than by a $\pi\pi$ resonance at 570 MeV.

E. The K(1400) meson

In an analysis of 3.5 GeV/c K$^-$ interactions on protons in a hydrogen bubble chamber, HAQUE et al. (1965) observed evidence for a Kπ resonance with mass and width given by

$$M = 1400 \pm 10 \text{ MeV}, \quad \Gamma \approx 160 \text{ MeV}.$$

The K (1400) appeared as a peak in the $K\pi$ mass spectrum in the reaction

$$K^- + p \rightarrow \overline{K}{}^0 + \pi^- + p .$$

The K (890) was also observed in this reaction, and indeed its production dominates the reaction.

The authors also looked at four-body final states in which an additional pion is produced. There was some slight evidence for the decay mode

$$K (1400) \rightarrow K (890) + \pi .$$

In this mode the maximum in the peak was at a mass of 1430 MeV.

To obtain information about the spin of the K(1400), the decay angular distribution was observed. The angular distribution was fitted in terms of density matrix elements described by GOTTFRIED and JACKSON (1964). Spin 2 is favored by the data, but spin 1 cannot be ruled out. Spin 0 is highly unlikely. Since the K(1400) decays strongly into a π and K, its parity is $P = (-1)^J$.

In a subsequent experiment with 3.9 and 4.2 GeV/c π^- mesons on protons, HARDY et al. (1965) found further evidence for the K (1400) at a mass and width

$$M = 1430 \pm 20 \text{ MeV}, \quad \Gamma = 100 \pm 20 \text{ MeV} .$$

These authors obtained evidence that the isospin is $I = 1/2$. Their analysis also favored a spin-parity assignment of $J^P = 2^+$, but $J^P = 1^-$ could not be ruled out. In summary, the most probable quantum numbers of the K (1400) are $I = 1/2$, $J^P = 2^+$.

F. Other possible $Y = 1$ meson states

In section 19 D, we discussed inconclusive evidence for one $K\pi\pi$ resonance at 1170 MeV and for another at 1230 MeV. MILLER et al. (1965), looking at π^-p interactions at 2.7 GeV/c, obtained additional evidence for the K (1170). However, neither this experiment nor the earlier one by WANGLER et al. (1964) gives very strong evidence for the existence of the K (1170) as a meson state. We caution that the sum of two inconclusive experiments does not necessarily equal a conclusive one.

In a study of $\overline{p}p$ annihilation at 3.0 GeV/c, BÖCK et al. (1964) obtained some evidence for still another $K\pi\pi$ resonance of mass and width

$$M = 1270 \text{ MeV}, \quad \Gamma = 60 \text{ MeV} .$$

The peak was found in looking at final states of $K\overline{K}\pi\pi\pi$. The $K\pi\pi$ was seen only in states with z-component of isospin $I_z = \pm 3/2$, but the authors state that peaks in the $I_z = \pm \, ^1/_2$ states might have been obscured by larger background. This state was observed to decay into K (890) $+ \pi$ and possibly also into $\rho + $ K. The existence of either of these decay modes (together with the observation $I_z = \, ^3/_2$) implies that the state has $I = \, ^3/_2$.

31. Recently discovered baryon states

A. Possible additional excited states of the nucleon

1. BAREYRE et al. (1964) have pointed out that there is a "shoulder" in the $I = \, ^1/_2$ pion-nucleon total cross section at an energy below the N(1510). If a Breit-Wigner resonance peak for the N(1510) is subtracted from the cross section, a peak appears at a mass of about 1400 MeV. Other experiments have also given evidence for a peak at about this mass, and phase shift analyses have indicated that the peak may be a πN resonance with $J^P = \, ^1/_2{}^+$. DALITZ and MOORHOUSE (1965) have shown that although the existing data are consistent with the assumption of such a resonance, they do not require it. Strong absorption occurs in this state, and this absorption can account for the peak. References to the experiments and analyses can be found in the paper of DALITZ and MOORHOUSE.

In an experiment with photons of energies up to 4 GeV incident on protons, R. ALVAREZ et al. (1964) have seen evidence for a number of excited nucleon states, including two previously unreported ones. The first of these has $M = 2550$ MeV and probably $I = \, ^3/_2$; the second has $M = 2700$ MeV and probably $I = \, ^1/_2$. Additional evidence is warrented before these states are accepted.

2*. Resonances or resonance-like phenomena have been detected at the following total c. m. energies (in MeV)

$T = 1/2$	1400?	1525	1680		2210		2640		3020	
	P_{11}	D_{13}	F_{15}							
$T = 3/2$	1236			1920		2420		2840		3220
	P_{33}			F_{37}						

In the case of P_{11} the energy refers to the maximum of $d \, \delta/dE$. The absorption parameter decreases rapidly to $\eta \approx 0.2$ at the kinetic energy of the pion $T_\pi = 600$ MeV.

---

* The paragraph A 2 was added by the editor. It contains some of the results presented in the invited papers of W. GALBRAITH, C. LOVELACE, R. G. MOORHOUSE and G. HÖHLER at the Royal Society meeting on pion-nucleon scattering and excited states of the nucleon (London, February 1965). The editor is indebted to Prof. CITRON, Prof. GALBRAITH, Dr. LOVELACE, Dr. MOORHOUSE and Dr. ROPER for private communications.

For N(1525) the real part of the phase shift $\delta = 90°$ at T_π $= 620$ MeV. This resonance is very inelastic ($\eta \approx 0.3$ at 620 MeV) and much narrower ($\delta = 45°$ at about $T_\pi = 580$ MeV) than formerly supposed.

The shoulder of the total $\pi^+ p$ cross section is not mentioned in the Table. An important contribution to this phenomenon belongs probably to S_{31}, which becomes almost completely absorptive near 800 MeV.

The peaks seen by ALVAREZ et al. (1964) in photoproduction are probably caused by $\Delta(2420)$ and N(2640) (HÖHLER et al. 1964).

The elasticity of several resonances was estimated from the variation of the real part of the forward amplitude. Assuming $\delta = 0°$ at resonance [cf. DALITZ (1963), p. 352] and the usual spin assignment, $x = \sigma^{el}/\sigma^{tot} \approx$ ≈ 0.35 for $\Delta(1920)$, $(J = 7/2)$ and $x \approx 0.18$ for N(2210), $(J = 9/2)$.

It is remarkable that most of the nucleon states [excluding N(1525) and the P_{11} phenomenon, but including N(938)] are almost equidistant in an s-scale [$s =$ total c. m. energy squared, see also Fig. 3a in the article of ROSENFELD (1964b)].

B. A possible $I = 2$ $\Sigma \pi$ resonance

PAN and ELY (1964), in looking at K^- interactions in carbon (from propane) at incident K momenta of 1.15 GeV/c, obtained some evidence for an $I = 2$ $\Sigma \pi$ resonance. The reaction is interpreted to be

$$K^- + n(C) \rightarrow (\Sigma^- \pi^-) + \pi^+ .$$

A peak in the $\Sigma^- \pi^-$ mass spectrum appeared at $M = 1415 \pm 16$ Mev. Because of a large background the width could not be determined. The evidence for this state is not convincing.

C. The Ξ (1820)

In looking at $K^- p$ interactions at incident K^- momenta between 2.45 and 2.70 GeV/c, SMITH et al. (1964, 1965) obtained good evidence for the existence of an excited Ξ state with mass and width

$$M = 1817 \pm 7 \text{ MeV} , \quad \Gamma = 60 \text{ MeV} .$$

Evidence for this state was also seen by BADIER et al. (1964). SMITH et al. obtained the following decay modes

$$\Xi(1820) \rightarrow \Lambda + \overline{K} \qquad (11.4 \pm 2.6)$$
$$\rightarrow \Xi + \pi \qquad (10.8 \pm 2.7)$$
$$\rightarrow \Xi(1530) + \pi \qquad (3.0 \pm 1.5)$$
$$\rightarrow \Xi + \pi + \pi \qquad (> 1.1)$$

where the numbers in parentheses are relative decay rates. These numbers also represent the cross sections in μb for the $\Xi(1820)$ to be produced and decay according to the various modes. BADIER et al. did not observe the decay into $\Xi + \pi$, obtaining an upper limit for production and decay into this mode of 1.2 ± 0.5 μb. The cause of this discrepancy is not yet understood.

The existence of the decay into $\Lambda + \overline{K}$ shows that the isospin of the state is $I = {}^1\!/_2$. Branching ratios into the various charge states in the other modes also favor this assignment. An analysis of the decay angular distribution is not conclusive, but weakly favors $J^P = {}^3\!/_2{}^-$. Insufficient statistics plus a large background make the determination uncertain. An assumption of spin ${}^5\!/_2$ was not necessary to fit the data but could not be excluded. The parity assignment was made assuming positive $\Xi\Lambda$ relative parity.

32. Recent measurements of particle masses, lifetimes, and decays

The data on the masses and other properties of the mesons and baryons appearing in sections 1 through 28 do not include results after March, 1964. Since that time, a number of more accurate measurements have been made. We give here some of the larger changes arising from the recent measurements. We also include the results of some measurements of quantities not previously determined. Unless the reference is given in the following, the data are taken from the compilation of ROSENFELD et al. (1964 a). References to the original papers may be found in this compilation.

A. Meson data

The partial decay rates of the K^+ and K_L in % are

$K^+ \to \mu^+ \nu$	63.1 ± 0.5	$K_L \to \pi^0 \pi^0 \pi^0$	27.1 ± 3.6
$\to \pi^+ \pi^0$	21.5 ± 0.4	$\to \pi^+ \pi^- \pi^0$	12.7 ± 1.7
$\to \pi^+ \pi^+ \pi^-$	5.5 ± 0.1	$\to \pi \mu \nu$	26.6 ± 3.2
$\to \pi^+ \pi^0 \pi^0$	1.7 ± 0.1	$\to \pi e \nu$	33.6 ± 3.3
$\to \pi^0 \mu^+ \nu$	3.4 ± 0.2	$\to \pi^+ \pi^-$	0.15 ± 0.03
$\to \pi^0 e^+ \nu$	4.8 ± 0.2		
$\to \pi^+ \pi^- e^+ \nu$	$(4.3 \pm 0.9) \times 10^{-3}$		
$\to \pi^+ \pi^- \mu^+ \nu \approx 10^{-3}$			
$\to \pi^+ \pi^+ e^- \bar{\nu} < 10^{-4}$			

The rate for $K^+ \to \pi^+ \pi^- \mu^+ \nu$ is based on one event (GREINER 1964). The $K_L \to 2\pi$ rate is from CHRISTENSON et al. (1964). The lifetime of K_L is $\tau_{K_L} = (5.6 \pm 0.7) \times 10^{-8}$ sec.

The partial decay rates of the η in % are

$\eta \to \gamma \gamma$	35.3 ± 3
$\to \pi^0 \pi^0 \pi^0$ (and $\pi^0 \gamma \gamma$)	31.8 ± 2.3
$\to \pi^+ \pi^- \pi^0$	27.4 ± 2.5
$\to \pi^+ \pi^- \gamma$	5.5 ± 1.3

The f has been observed (GELFAND 1964) to decay into $2\pi^0$ with the branching fraction

$$\frac{f \to \pi^0 \pi^0}{f \to \pi^+ \pi^-} = 0.6 \pm 0.17 .$$

This is consistent with the assignment $I = 0$ for the f, and shows that the f cannot be the same as the B.

B. Baryon data

The Λ decays into the mode $p\, e^- \nu$ with a branching fraction $(0.88 \pm 0.08) \times 10^{-3}$[*].

The masses of the Σ hyperons are

$$M_{\Sigma^+} = 1189.41 \pm 0.14 \text{ MeV}$$
$$M_{\Sigma^0} = 1192.3 \pm 0.3$$
$$M_{\Sigma^-} = 1197.08 \pm 0.19 .$$

With these values of the masses, the mass differences are

$$M_{\Sigma^0} - M_{\Sigma^+} = 2.9 \pm 0.3 , \quad M_{\Sigma^-} - M_{\Sigma^0} = 4.75 \pm 0.1 .$$

Previously, these mass differences were equal within experimental error. A measurement of the magnetic moment of the Σ^+ has been made (McINTURFF 1964). The value is

$$\mu_{\Sigma^+} = 4.3 \pm 1.5 \quad \text{nuclear Bohr magnetons} .$$

The partial decay rates (fraction of total) of the Σ^+ and Σ^- into rare modes are

$\Sigma^+ \to n \pi^+ \gamma$	0.4×10^{-4}	$\Sigma^- \to n \pi^- \gamma$	0.1×10^{-4}
$\to \Lambda e^+ \nu$	0.2×10^{-4}	$\to n \mu^- \nu$	$(0.66 \pm 0.14) \times 10^{-3}$
$\to p \gamma$	$(0.19 \pm 0.04) \times 10^{-2}$	$\to n e^- \nu$	$(1.4 \pm 0.3) \times 10^{-3}$
$\to n \mu^+ \nu$	$< 2.3 \times 10^{-4}$	$\to \Lambda e^- \nu$	$(0.75 \pm 0.28) \times 10^{-4}$
$\to n e^+ \nu$	$< 1.0 \times 10^{-4}$		

The rate for $\Sigma^+ \to p \gamma$ is from BAZIN et al. (1965).

---

[*] CHARRIÈRE et al. (1965) have determined a new value of the magnetic moment of Λ: $\mu_\Lambda = -0.5 \pm 0.28$ nuclear magnetons (added by the editor).

The mean lives of the Ξ^- and Ξ^0 are

$$\tau_{\Xi^-} = (1.74 \pm 0.05) \times 10^{-10} \text{ sec}$$
$$\tau_{\Xi^0} = (3.06 \pm 0.40) \times 10^{-10} \text{ sec} .$$

A rate for leptonic Ξ^- decay has been measured. The decay mode and branching fraction are

$$\Xi^- \to \Lambda \, e^- \, \nu \quad (3.0 \pm 1.7) \times 10^{-3} .$$

At least four Ω^- particles have been seen. The mass is based on the first two events; the mean life (ABRAMS 1964) is based on four events.

$$M = 1675 \pm 3 \text{ MeV}$$
$$\tau = (1.3 \pm 0.7) \times 10^{-10} \text{ sec} .$$

In addition to the $\Xi^0 \, \pi^-$ and ΛK^- decay modes given in section 27a, the decay

$$\Omega^- \to \Xi^- + \pi^0$$

has been seen.

Electromagnetic mass splittings among the members of baryon excited states have been reported by HUWE (1964), COOPER et al. (1964), OLSSON (1965), and PJERROU et al. (1965). References to attempts at theoretical interpretation according to unitary symmetry can be found in these papers. Values of masses and width in MeV are

$$M_{\Delta^{++}} = 1236.0 \pm 0.5 , \quad M_{\Delta^0} = 1236.45 \pm 0.65$$
$$M_{\Delta^{++}} - M_{\Delta^0} = -0.45 \pm 0.85$$
$$\Gamma_{\Delta^{++}} = 120.0 \pm 2.0 , \quad \Gamma_{\Delta^0} = 119.6 \pm 2.4$$
$$\Gamma_{\Delta^{++}} - \Gamma_{\Delta^0} = 0.4 \pm 3.1$$
$$M_{\Sigma^{*-}} - M_{\Sigma^{*+}} = 5.2 \pm 2.2$$
$$M_{\Xi^{*-}} - M_{\Xi^{*0}} = 6.8 \pm 3.3 .$$

The values of the masses of resonances can vary from experiment to experiment, so perhaps the quoted errors, which are based on individual experiments, should not be taken too seriously.

The quasi two-body decay $\Sigma(1660) \to \Lambda(1450) + \pi$ has been seen (EBERHARD 1965) with the branching ratio.

$$\frac{\Sigma(1660) \to \Lambda(1405) + \pi}{\Sigma(1660) \to \text{all } \Sigma \pi \pi} = 0.9 \begin{array}{c} +0.10 \\ -0.16 \end{array}$$

C. Lower limits on masses of conjectured particles

1. Although a boson $W^\pm$ to mediate the weak interactions has been looked for in several experiments, it has not been found. BLOCK et al. (1964) put a lower limit on the mass of the W: $M_W > 1.5$ GeV. Assuming

that the W decays leptonically at least half the time, BERNARDINI et al. (1964) were able to obtain a somewhat higher lower limit: $M_W > 1.8\,\text{GeV}$. SUNDERLAND et al. (1965) obtained a lower limit $M_W > 2\,\text{GeV}$, again assuming at least a 50% branching ratio into leptons in the decay.

2. Certain unitary symmetry models of elementary particles (to be discussed briefly in the next section) require the existence of heavy particles of mass much greater than 1 GeV. These particles are supposed to be either integrally or fractionally charged in units of the electron charge, depending on the model. Such particles have been looked for both at accelerators and in cosmic rays, and so far have not been found. One can give a lower limit on the mass only by assuming a particular value for the production cross section. Results from accelerators (MORRISON 1964, BINGHAM 1964, LEIPUNER 1964, FRANZINI 1965) give lower limits of between two and three GeV for "reasonable" strong interaction production cross sections. (Not all the references are given; for some others see FRANZINI 1965). BOWEN et al. (1964) have looked for fractionally charged particles in a cosmic ray experiment, obtaining a lower limit on the mass of ≈ 10 to $\approx 15\,\text{GeV}$, depending on the assumed production cross section. Other cosmic ray experiments are in progress.

33. Classification according to SU_6

A. Quark Model

The classification scheme discussed in section 28 is based on the 8 dimensional representation of SU_3. It is a puzzle why particles have not been seen which correspond to the two inequivalent 3 dimensional representations (called fundamental representations) of SU_3.

To understand why we expect the fundamental representations to occur, we make an analogy between SU_3 and SU_2. The representations (including double-valued representations) of the group of rotations in ordinary space or in isospin space are the same as the representations of SU_2. It is therefore convenient to regard the spin and isospin groups as being SU_2 groups. Now the (one) fundamental representation of SU_2 is two-dimensional and corresponds to a particle of spin or isospin $1/2$. Particles of spin or isospin $1/2$ occur in nature, of course, a simple example being the nucleon which is an SU_2 doublet with respect to both spin and isospin. However in SU_3 the present situation seems analogous to what nature would be like in SU_2 if only integral spins (or isospins) were realized.

GELL-MANN (1964) and ZWEIG (1964) have postulated a model in which the mesons and baryons are constructed of triplets of particles (called quarks or aces) which correspond to the fundamental representations of SU_3. The mesons are constructed from a product of quark and antiquark according to the series

$$3 \times \overline{3} = 1 + 8 .$$

Since mesons are observed in singlets and octuplets, this scheme gives the wanted representations, while dispensing with the representation 10, $\overline{10}$ and 27 which occur in the product 8×8 but which do not correspond to known mesons. The baryons are constructed from a product of three quarks according to the scheme:

$$3 \times 3 \times 3 = 1 + 8 + 8 + 10 .$$

Again the representations $\overline{10}$ and 27, for which there are no known baryons, are missing.

One interesting feature of the quark scheme is that if the baryons are to have integral charge, hypercharge and baryon number, the quarks must have fractional values of these quantum numbers. In particular quarks must have baryon number $B = {}^1/_3$ and Q and Y given by $- {}^1/_3$ and $+ {}^2/_3$. Although in the model the quark building blocks for the mesons and baryons need not be realized, there is at least the possibility that they might correspond to actual particles. However, as discussed in the previous section, fractionally charged particles have been looked for, so far without success.

B. Integrally charged triplets

To avoid the "unpleasant" feature of particles with fractional charge, a number of authors have suggested models with particles of integral charge, hypercharge, and baryon number. Such models require at least two different kinds of triplets. Rather than give all the references, we merely mention the paper of GÜRSEY, LEE, and NAUENBERG (1964) which contains a large number of possible models. A model with two kinds of triplets has an advantage of economy over a model with three kinds of triplets. However, in the latter model, all three kinds of triplets are on an equal footing in the product $3 \times 3 \times 3$. Massive particles of integral charge have been looked for but not found.

C. SU_6 symmetry

In either a model with quarks or integrally charged triplets, these particles must have half-integral spin; otherwise baryons would not have half-integral spin. The simplest possibility, as far as the spin degree of freedom is concerned, is to let the triplets belong to the fundamental two dimensional representation of SU_2 (i. e. have spin ${}^1/_2$). Thus, the symmetry of a triplet is the product symmetry $SU_3 \times SU_2$ (SU_3 for internal symmetry, SU_2 for spin). Considered in this way, a triplet has six degrees of freedom, since each member can have spin up or spin down. SAKITA (1964a) and GÜRSEY and RADICATI (1964b) postulated the existence of a still greater symmetry SU_6, which corresponds to symmetry under transformations which mix spin and internal degrees of freedom. This symmetry was proposed in analogy with the SU_4 symmetry proposed by WIGNER (1937) to include together the spin and isospin degrees of freedom of the nucleon.

In SU_6, the mesons are obtained from the product

$$6 \times \overline{6} = 1 + 35$$

likewise the baryons are obtained from the product

$$6 \times 6 \times 6 = 20 + 56 + 70 + 70 .$$

In order to compare these numbers within the known mesons and baryons, we must give their $SU_3 \times SU_2$ content. For the mesons this is

$$35 = 8 \cdot 1 + 8 \cdot 3 + 1 \cdot 3$$

$$1 = 1 \cdot 1 .$$

The notation $8 \cdot 3$ means an octuplet of particles, each of which is a spin triplet, i. e. an octuplet of particles of spin 1. We can now identify the 35 dimensional representation as containing the eight pseudoscalar mesons ($\pi \, \eta \, K$) in $8 \cdot 1$ and the nine pseudovector mesons ($\rho \, \omega \, K^* \, \varphi$) in $8 \cdot 3 + 1 \cdot 3$. The pseudoscalar η' meson is the singlet $1 \cdot 1$.

Thus, four SU_3 multiplets are contained in two SU_6 multiplets. Also, the 9 mesons of spin 1 and the 8 of spin 0 have the same parity, as they must if they are to belong to the same multiplet. Furthermore, according to the theory, the ω and φ are degenerate in mass. Consequently, when the symmetry is broken, their mutual effect on each other spoils the GELL-MANN-OKUBO mass formula. On the other hand, the pseudoscalar η' does not belong to the same representation as the η and therefore needs not be degenerate with it. Therefore, the effect of the η' does not spoil the mass formula for the pseudoscalar octuplet.

Among the baryons, the 20, 56, and 70 dimensional representations have the $SU_3 \times SU_2$ content

$$20 = 1 \cdot 4 + 8 \cdot 2$$

$$56 = 8 \cdot 2 + 10 \cdot 4$$

$$70 = 1 \cdot 2 + 8 \cdot 2 + 8 \cdot 4 + 10 \cdot 2 .$$

The 8 spin $^1/_2$ baryons ($N \, \Lambda \, \Sigma \, \Xi$) and the 10 spin $^3/_2$ baryons ($\Delta \, \Sigma^* \, \Xi^* \, \Omega$) can be fitted very nicely into the 56 multiplet. A mass formula relates the average splitting of the octuplet to the splitting of the decuplet.

It is not yet clear how to classify the other mesons and baryons, perhaps because not all their quantum numbers are known, and perhaps because unitary symmetry fails for these higher energy states.

Electromagnetic effects can also be treated. We merely mention the interesting prediction that the ratio of the magnetic moment of the neutron to that of the proton is given by (SAKITA 1964b, BÉG 1964)

$$\mu_n/\mu_p = -2/3 \, .$$

The experimental number is -0.685.

However despite these successes of SU_6, the scheme suffers from at least one major difficulty. This is that the theory is non-relativistic, as can be seen by the following argument: Although the symmetry mixes spin and internal degrees of freedom, it does not mix internal symmetry with orbital angular momentum. But the requirements of special relativity of necessity mix spin and orbital angular momentum, and thereby clash with SU_6. We shall not list the many attempts, none wholly successful, to overcome this difficulty by doubling the number of degrees of freedom to 12. At the very least SU_6 provides for a simple classification of the low-lying meson and baryon states.

Acknowledgements

During the preparation of this work I have had beneficial discussions with many physicists, especially J. BALLAM, S. M. BERMAN, S. GLASHOW, H. P. NOYES, M. PESHKIN, A. H. ROSENFELD, M. H. ROSS, M. T. VAUGHN, R. J. OAKES, A. ODIAN, M. PERL, M. WHITEHEAD, and C. N. YANG. I am also grateful to the many physicists who have sent me preprints of their work or who have permitted me to use figures or tables from their published papers. Work done on the previously unpublished material appearing in these notes was supported in part by the U.S. Atomic Energy Commission and by the U. S. National Science Foundation. I should like to thank Professor W. K. F. PANOFSKY for his hospitality during my stay at the Stanford Linear Accelerator Center. Finally, I am grateful to the Physics Division of the Aspen Institute for Humanistic Studies, where I spent part of the summer of 1963 in the quiet, reflective atmosphere of the Rocky Mountains. It is a pleasure to thank Professor G. HÖHLER for his helpful comments and encouragement.

References for sections 1—28 *

ABASHIAN, A., N. BOOTH, and K. CROWE: Phys. Rev. Letters 5, 258 (1960)
— — —, R. HILL, and E. ROGERS: Phys. Rev. 132, 2296 (1963)
ABOLINS, M. R. LANDER, W. MEHLHOP, N. XUONG, and P. YAGER: Phys. Rev. Letters 11, 381 (1963)
ADAIR, R. K.: Phys. Rev. 100, 1540 (1955)
—, and E. C. FOWLER: Strange Particles. New York: John Wiley 1963
ADEMOLLO, M., and R. GATTO: Phys. Rev. 133, B 531 (1964)
ALEXANDER, G., O. DAHL, L. JACOBS, G. KALBFLEISCH, D. MILLER, A. RITTEN-BERG, J. SCHWARTZ, and G. SMITH: Phys. Rev. Letters 9, 460 (1962)
—, L. JACOBS, G. KALBFLEISCH, D. MILLER, G. SMITH, and J. SCHWARTZ: CERN (1962a), p. 320
—, G. KALBFLEISCH, D. MILLER, and S. SMITH: Phys. Rev. Letters 8, 447 (1962b)
ALFF, C., D. BERLEY, D. COLLEY, N. GELFAND, U. NAUENBERG, D. MILLER, J. SCHULTZ, J. STEINBERGER, T. TAN, H. BRUGGER, P. KRAMER, and R. PLANO: Phys. Rev. Letters 9, 322 (1962); 9, 325 (1962)
ALSTON, M., L. ALVAREZ, P. EBERHARD, M. GOOD, W. GRAZIANO, H. TICHO, and S. WOJCICKI: Phys. Rev. Letters 5, 520 (1960)
— — — — — — — Phys. Rev. Letters 6, 300 (1961)
— — — — — — — Phys. Rev. Letters 6, 698 (1961a)
— —, M. FERRO-LUZZI, A. ROSENFELD, H. TICHO, and S. WOJCICKI: CERN (1962), p. 311
ALVAREZ, L., M. ALSTON, M. FERRO-LUZZI, D. HUWE, G. KALBFLEISCH, D. MILLER, J. MURRAY, A. ROSENFELD, J. SHAFER, F. SOLMITZ, and S. WOJCICKI: Phys. Rev. Letters 10, 184 (1963)
ARMENTEROS, R., D. EDWARDS, T. JACOBSEN, A. SHAPIRA, J. VANDERMEULEN, C. D'ANDLAU, A. ASTIER, P. BAILLON, H. BRIAND, J. COHEN-GANOUNA, C. DE-FOIX, J. SIAUD, C. GHESGUIÈRE, and P. RIVET: Siena (1963)
— — — — — — — —, J. COHEN-GANOUNA, C. DEFOIX, J. SIAUD, C. GHES-GUIÈRE, and P. RIVET: CERN preprint (1964)
— L. MONTANET, D. MORRISON, S. NILSSON, A. SHAPIRA, J. VAN-DERMEULEN, C. D'ANDLAU, A. ASTIER, C, GHESGUIÈRE, B. GREGORY, D. RAHM, P. RIVET, and F. SOLMITZ: CERN (1962), p. 295
BACASTOW, R., T. ELIOFF, R. LARSEN, C. WIEGAND, and T. YPSILANTIS: Phys. Rev. Letters 9, 400 (1962)
BALTAY, C., J. SANDWEISS, H. TAFT, B. CULWICK, W. FOWLER, J. KOPP, R. LOUTTIT, J. SANFORD, R. SHUTT, D. STONEHILL, A. THORNDIKE, and M. WEBSTER: Phys. Rev. Letters 11, 346 (1963)
BARBARO-GALTIERI, A., A. HUSSAIN, and R. D. TRIPP: Phys. Letters 6, 296 (1963).
BAREYRE, P., C. BRICMAN, G. VALLADAS, G. VILLET, J. BIZARD, J. SEGUINOT: Physics Lett. 8, 137 (1964)
BARKAS, W. H., and A. H. ROSENFELD: University of California Report UCRL 8030 (revised 1963)
BARLOUTAUD, R., J. HEUGHEBAERT, A. LEVEQUE, J. MEYER, and R. OMNES: Phys. Rev. Letters 8, 32 (1962)
BARNES, V. E., et al.: Phys. Rev. Letters 12, 204 (1964)
BARRET, B., and G. BARTON: Phys. Rev. 133, B 466 (1964).
BARSHAY, S.: Phys. Letters 3, 320 (1963)
BARTON, G., and S. P. ROSEN: Phys. Rev. Letters 8, 414 (1962).
BASTIEN, P., and J. BERGE: Phys. Rev. Letters 10, 188 (1963).

---

* Three non-standard abbreviations are used in these references:

1. "CERN (1962)" for the 1962 International Conference on High-Energy Physics at CERN, (Geneva), Proceedings.

2. "Stanford (1963)" for the International Conference on Nucleon Structure, Stanford University, June, 1963 (Proceedings to be published).

3. "Siena (1963)" for the Proceedings of the Sienna International Conference on Elementary Particles (1963).

BASTIEN, P., J. BERGE, O. DAHL, M. FERRO-LUZZI, W. HUMPHREY, J. KIRZ, D. MILLER, J. MURRAY, A. ROSENFELD, M. ROSS, J. SCHWARTZ, F. SOLMITZ, R. TRIPP, and M. WATSON: (quoted in Chamberlain 1962)
— — — —, D. MILLER, J. MURRAY, A. ROSENFELD, and M. WATSON: Phys. Rev. Letters 8, 114 (1962a)
— — — —, J. KIRZ, D. MILLER, J. MURRAY, A. ROSENFELD, R. TRIPP, and M. WATSON: CERN (1962b), p. 373
—, M. FERRO-LUZZI, and A. H. ROSENFELD: Phys. Rev. Letters 6, 702 (1961)
BEALL, E., W. HOLLEY, D. KEEFE, L. KERTH, J. THRESHER, C. WANG, and W. WENZEL: CERN (1962), p. 368
BÉG, M. A. B.: Phys. Rev. Letters 9, 67 (1962)
BEHRENDS, R., J. DREITLEIN, C. FRONSDAL, and B. LEE: Rev. Mod. Phys. 34, 1 (1962)
BELYAKOV, V., WANG YUNG-CHANG, V. VEKSLER, N. VIRYASOV, DU YUAN-CAI, E. KLADNITSKAYA, KIM HI IN, A. KUZNETSOV, A. MIKHUL, NGUYEN DINH-TU, V. PENEV, E. SOKOLOVA, and M. SOLOV'EV: CERN (1962), p. 336
BERMAN, S. M., and S. D. DRELL: Phys. Rev. Letters 11, 220 (1963).
BERTANZA, L., V. BRISSON, P. CONNOLLY, E. HART, I. MITTRA, G. MONETI, R. RAU, N. SAMIOS, S. LICHTMAN, I. SKILLICORN, L. GRAY, M. GOLDBERG, J. LEITNER, and J. WESTGARD: Phys. Rev. Letters 9, 180 (1962a)
—, P. CONNOLLY, B. CULWICK, F. EISLER, T. MORRIS, R. PALMER, A. PRODELL, and N. SAMIOS: Phys. Rev. Letters 8, 332 (1962)
BETHE, H. A., and F. DE HOFFMANN: Mesons and Fields. Vol. 2. Evanston, Illinois: Row, Peterson, and Company 1955
BIRGE, R., R. ELY, G. GIDAL, G. KALMUS, A. KERNAN, W. POWELL, U. CAMERINI, W. FRY, J. GAIDOS, R. MARCH, and S. NATALI: Phys. Rev. Letters 11, 35 (1963)
— —, W. POWELL, H. HUZITA, W. FRY, J. GAIDOS, S. NATALI, R. WILLMAN, and U. CAMERINI: Proc. of the 1960 Annual International Conference on High Energy Physics at Rochester (1960), p. 601
BISI, V., G. BORREANI, R. CESTER, A. DEBENEDETTI, M. FERRERO, C. GARELLI, A. MARZARI-CHIESA, B. QUASSIATI, G. RINAUDO, A. TRABUCCO, M. VIGONE, and A. WERBROUCK: Phys. Rev. Letters 10, 498 (1963)
BLATT, J. M., and V. F. WEISSKOPF: Theoretical Nuclear Physics. New York: John Wiley and Sons 1952
BLOCK, M., T. KIKUCHI, D. KOETKE, J. KOPELMAN, C. SUN, R. WALKER, G. CULLIGAN, V. TELEGDI, and R. WINSTON: Phys. Rev. Letters 11, 301 (1963)
—, L. LENDINARA, and L. MONARI: CERN (1962), p. 371
BOGOLIUBOV, N. N., and D. SHIRKOV: Introduction to the Theory of Quantized Fields. New York: Interscience Publishers 1959
BOHR, A.: Nuclear Phys. 10, 486 (1959)
BONDÀR, L., E. KEPPEL, G. KRAUS, W. DODD, B. TALLINI, G. WOLF, I. BUTTERWORTH, F. CAMPAYNE, M. IBBOTSON, N. BISWAS, I. DERADO, D. LUERS, and N. SCHMITZ: Phys. Letters 5, 209 (1963)
BOOTH, N. E.: Phys. Rev. 132, 2305 (1963)
—, and A. ABASHIAN: Phys. Rev. 132, 2314 (1963)
— —, and K. M. CROWE: Phys. Rev. Letters 7, 35 (1961)
— — — Phys. Rev. 132, 2309 (1963)
BRONZAN, J. B., and F. E. LOW: Phys. Rev. Letters 12, 522 (1964)
BROWN, L. M., and P. SINGER: Phys. Rev. Letters 8, 460 (1962)
— — Phys. Rev. 133, B 812 (1964).
BRUECKNER, K., R. SERBER, and K. WATSON: Phys. Rev. 81, 575 (1951)
BURGOYNE, N.: Nuovo cimento 8, 607 (1958)   ·
BURHOP, E. H. S.: The Auger Effect. Cambridge: Cambridge University Press 1952
BYERS, N., and S. FENSTER: Phys. Rev. Letters 11, 52 (1963)
CALDWELL, D. O.: Phys. Rev. Letters 7, 259 (1961)
—, E. BLEULER, L. JONES, B. ELSNER, W. MIDDLECOOP, D. HARTUNG, and B. ZACHAROW: New York Meeting of the American Physical Society (January, 1964), post deadline paper
CAMERINI, U., W. F. FRY and J. GAIDOS: Nuovo cimento 28, 1096 (1963)

CARMONY, D., R. L. LANDER, C. RINDFLEISCH, N. XUONG, and P. YAGER: Phys. Rev. Letters 12, 254 (1964).
—, A. ROSENFELD, and R. VAN DE WALLE: Phys. Rev. Letters 8, 117 (1962)
CARMONY, D. D., G. M. PJERROU, P. E. SCHLEIN, W. E. SLATER, D. H. STORK, and H. K. TICHO: Phys. Rev. Letters 12, 482 (1964)
CARRUTHERS, P.: Phys. Rev. 133, B 497 (1964).
CASON, N., M. GOOD, and W. WALKER: to be published (1964)
CENCE, R., T. DEVLIN, R. EANDI, D. HAGGE, J. HELLAND, R. KENNEY, P. MCMANI-GAL, B. MOYER, P. OGDEN, and V. PETERSON: Stanford (1963)
CHAMBERLAIN, O., K. CROWE, D. KEEFE, L. KERTH, A. LEMONICK, TIN MAUNG, and T. ZIPF: Phys. Rev. 125, 1696 (1962)
CHAN, C. H., Physics Lett. 8, 211 (1964).
CHEN, K., A. CONE, J. DUNNING JR., S. FRANK, N. RAMSEY, J. WALKER, and R. WILSON: Phys. Rev. Letters 11, 561 (1963)
CHEW, G. F.: S-Matrix Theory of Strong Interactions. New York: W. A. Benjamin, Inc., 1961
—, and S. C. FRAUTSCHI: Phys. Rev. Letters 7, 394 (1961)
— — Phys. Rev. Letters 8, 41 (1962)
—, and F. E. LOW: Phys. Rev. 113, 1640 (1959)
—, and G. C. WICK: Phys. Rev. 85, 636 (1952)
CHINOWSKY, W., G. GOLDHABER, S. GOLDHABER, W. LEE, and T. O'HALLORAN: Phys. Rev. Letters 9, 330 (1962)
—, and J. STEINBERGER: Phys. Rev. 95, 1561 (1954)
CHRÉTIEN, M., F. BULOS, H. CROUCH, R. LANOU, J. MASSIMO, A. SHAPIRO, J. AVE-RELL, C. BORDNER JR., A. BRENNER, D. FIRTH, M. LAW, E. RONAT, K. STRAUCH, J. STREET, J. SZYMANSKI, A. WEINBERG, B. NELSON, I. PLESS, L. ROSENSON, G. SALANDIN, R. YAMAMOTO, L. GUERRIERO, and F. WALDNER: Phys. Rev. Letters 9, 127 (1962)
COHEN, E. R., and J. W. M. DUMOND: Report to the Commission on Nuclidic Masses and Related Atomic Constants of the I. U. P. A. P. (June 24, 1963)
COLLEY, D., N. GELFAND, U. NAUENBERG, J. STEINBERGER, S. WOLF, H. BRUGGER, P. KRAMER, and R. PLANO: CERN (1962), p. 315; Phys. Rev. 128, 1930 (1962)
CONNOLLY, P., E. HART, G. KALBFLEISCH, K. LAI, G. LONDON, G. MONETI, R. RAU, N. SAMIOS, I. SKILLICORN, S. YAMAMOTO, M. GOLDBERG, M. GUNDZIK, J. LEITNER, and S. LICHTMAN: Siena (1963)
— —, K. LAI, G. LONDON, G. MONETI, R. RAU, N. SAMIOS, I. SKILLICORN, S. YAMA-MOTO, M. GOLDBERG, M. GUNDZIK, J. LEITNER, and S. LICHTMAN: Phys. Rev. Letters 10, 371 (1963a); Siena (1963a)
COOK, V., B. CORK, T. HOANG, D. KEEFE, L. KERTH, W. WENZEL, and T. ZIPF: Phys. Rev. 123, 320 (1961)
COOPER, W., H. COURANT, H. FILTHUTH, E. MALAMUD, A. MINGUZZI-RANZI, H. SCHNEIDER, A. SEGAR, G. SNOW, W. WILLIS, E. GELSEMA, J. KLUYVER, A. TENNER, K. BROWNING, I. HUGHES, and R. TURNBULL: CERN (1962), p. 298
COURANT, H., H. FILTHUTH, P. FRANZINI, R. GLASSER, A. MINGUZZI-RANZI, A. SEGAR, W. WILLIS, R. BURNSTEIN, T. DAY, B. KEHOE, A. HERZ, M. SAKITT, B. SECHI-ZORN, N. SEEMAN, and G. SNOW: Phys. Rev. Letters 10, 409 (1963)
CRAWFORD JR., F., R. GROSSMAN, L. LLOYD, L. PRICE, and E. FOWLER: Phys. Rev. Letters 11, 564 (1963a)
— L. LLOYD, and E. C. FOWLER: Phys. Rev. Letters 10, 546 (1963)
CRONIN, J. W., and O. E. OVERSETH: Phys. Rev. 129, 1795 (1963)
DALITZ, R. H.: Proc. Phys. Soc. (London) A 64, 667 (1951)
— Phil. Mag. 44, 1068 (1953)
— Rev. Mod. Phys. 31, 823 (1959)
— Ann. Rev. Nuclear Sci. 13, 339 (1963)
—, and S. TUAN: Ann. Physics 10, 307 (1960)
DANYSZ, M., and J. PNIEWSKI: Phil. Mag. 44, 348 (1953)
DAY, T., G. SNOW, and J. SUCHER: Phys. Rev. Letters 3, 61 (1959)
DEPOMMIER, P., J. HEINTZE, C. RUBBIA, and V. SOERGEL: Phys. Letters 5, 61 (1963)
D'ESPAGNAT, B.: Nuovo cimento 20, 1217 (1961)
DEVLIN, T. J., B. J. MOYER, and V. PEREZ-MENDEZ: Phys. Rev. 125, 690 (1962)

DIDDENS, A., E. JENKINS, T. KYCIA, and K. RILEY: Phys. Rev. Letters 10, 262 (1963)
— — — — Phys. Rev. 132, 2721 (1963a)
DOEDE, J., R. HILDEBRAND, M. ISRAEL, and M. PYKA: Phys. Rev. 129, 2808 (1963)
DRELL, S. D., and F. ZACHARIASEN: Electromagnetic Structure of Nucleons. Oxford: Oxford University Press 1961
DUNAITSEV, A. F., V. I. PETRUKHIN, YU. D. PROKSHKIN, and V. I. PYKALIN: Phys. Letters 1, 138 (1962)
EBERHARD, P., and M. L. GOOD: Phys. Rev. 120, 1442 (1960)
EDMONDS, A. R.: Angular Momentum in Quantum Mechanics. Princeton: Princeton University Press 1957
EISLER, F., P. FRANZINI, J. GAILLARD, A. GARFINKEL, J. KEREN, R. PLANO, A. PRODELL, and M. SCHWARTZ: Rev. Mod. Phys. 33, 436 (1961)
ELY, R., SUN-YUI FUNG, G. GIDAL, YU-LI PAN, W. POWELL, and H. WHITE: Phys. Rev. Letters 7, 461 (1961)
EVANS, L. E.: Nuovo cimento 25, 580 (1962) and University of Wisconsin preprint (unpublished)
ERWIN, A., G. HOYER, R. MARCH, W. WALKER, and T. WANGLER: Phys. Rev. Letters 9, 34 (1962)
—, R. MARCH, and W. WALKER: Nuovo cimento 24, 237 (1962a); Phys. Letters 3, 99 (1962a)
— — —, and E. WEST: Phys. Rev. Letters 6, 628 (1961)
FABRI, E.: Nuovo cimento Suppl. 12, 205 (1954)
FANO, U.: Rev. Mod. Phys. 29, 74 (1957)
FEINBERG, G.: Phys. Rev. 109, 1019 (1958)
—, and R. E. BEHRENDS: Phys. Rev. 115, 745 (1959)
FELD, B. T., and W. M. LAYSON: CERN (1962), p. 147
—, and D. B. LICHTENBERG: Nuovo cimento 22, 996 (1961)
FELDMAN, G., and T. FULTON: Nuclear Phys. 8, 106 (1958)
FERRETTI, B.: Report of an International Conference on Fundamental Particles and Low Temperatures. Cambridge (1946), 1, 75, Physical Society. London 1947
FERRO-LUZZI, M., F. SOLMITZ, and M. STEVENSON: CERN (1962), p. 376
—, R. TRIPP, and M. WATSON: Phys. Rev. Letters 8, 28 (1962a)
FEYNMAN, R. P.: Quantum Electrodynamics. New York: W. A. Benjamin 1961
— Theory of Fundamental Processes. New York: W. A. Benjamin 1962
—, and M. GELL-MANN: Phys. Rev. 109, 193 (1958)
FETKOVICH, J. G., and E. G. PEWITT: Phys. Rev. Letters 11, 290 (1963)
FIELDS, T. H., G. B. YODH, M. DERRICK, and J. G. FETKOVICH: Phys. Rev. Letters 5, 69 (1960)
FUELSCHE, H., E. FOWLER, H. KRAYBILL, J. SANFORD, and D. STONEHILL: Phys. Rev. Letters 9, 223 (1962)
FOLDY, L. L.: Phys. Rev. 102, 568 (1956)
FRANZINETTI, C., and G. MORPURGO: Nuovo cimento Suppl. 6, 469 (1957)
FRAUTSCHI, S. C.: Regge Poles and S-Matrix Theory. New York: W. A. Benjamin 1964
FRAZER, W. R., and J. R. FULCO: Phys. Rev. Letters 2, 365 (1959)
—, S. H. PATIL, and N. XUONG: Phys. Rev. Letters 12, 178 (1964)
FRISK, A., and A. EKSPONG: Phys. Letters 3, 27 (1962)
FULTON, T., G. KALLEN, J. D. JACKSON, and C. FRONSDAL: Elementary Particle Physics and Field Theory. New York: W. A. Benjamin 1963
GELFAND, N., D. MILLER, M. NUSSBAUM, J. RATAU, J. SCHULTZ, J. STEINBERGER, T. H. TAN, L. KIRSCH, and R. PLANO: Phys. Rev. Letters 11, 436 (1963); 11, 438 (1963a)
GELL-MANN, M.: Phys. Rev. 92, 833 (1953)
— Nuovo cimento Suppl. 2, 848 (1956)
— Phys. Rev. 106, 1296 (1957)
— Symmetries of Baryons and Mesons. California Institute of Technology Report CTSL-20 (1961) (unpublished)
— Phys. Rev. 125, 1067 (1962)
— CERN (1962b), p. 805
—, and A. PAIS: Proc. of the International Conference on High Energy Physics, Glasgow. London: Pergamon Press 1955

154

GELL-MANN, M., and A. PAIS: Phys. Rev. **97**, 1387 (1955a)
—, and A. H. ROSENFELD: Ann. Rev. Nuclear Sci. **7**, 407 (1957a)
—, D. SHARP, and W. WAGNER: Phys. Rev. Letters **8**, 261 (1962a)
GENTILE, G.: Nuovo cimento **17**, 493 (1940)
GERSTEIN, S., S., and YA. B. ZELDOVICH: JEPT **29**, 698 (1955) [Soviet Physics: JETP **2**, 576 (1956)]
GLASHOW, S.: Proc. Athens Topical Conf. on Recently Discovered Resonant Particles, Ohio Univ. Athens Ohio (1963)
—, and A. ROSENFELD: Phys. Rev. Letters **10**, 192 (1963)
—, and J. J. SAKURAI: Nuovo cimento **25**, 337 (1962); **26**, 622 (1962)
GLAUBER, R. J.: Phys. Rev. **100**, 242 (1955)
GOEBEL, C.: Phys. Rev. **103**, 258 (1956)
GOLDBERG, M., M. GUNDZIK, S. LICHTMAN, J. LEITNER, M. PRIMER, P. L. CONNOLLY, E. L. HART, K. W. LAI, G. LONDON, N. P. SAMIOS, and S. S. YAMAMOTO: Phys. Rev. Letters **12**, 546 (1964)
GOLDBERGER, M. L., and S. B. TREIMAN: Phys. Rev. **110**, 1178 (1958)
— — Nuovo cimento **9**, 451 (1958a)
GOLDHABER, G., J. BROWN, S. GOLDHABER, J. KADYK, B. SHEN, and G. TRILLING: Phys. Rev. Letters **12**, 336 (1964)
—, W. CHINOWSKY, S. GOLDHABER, W. LEE, and T. O. HALLORAN: Phys. Letters **6**, 62 (1963)
GOOD, R., R. MATSEN, F. MULLER, O. PICCIONI, W. POWELL, H. WHITE, W. FOWLER, and R. BIRGE: Phys. Rev. **124**, 1223 (1961)
GOURDIN, M.: this volume, page 1
GREEN, H. S.: Phys. Rev. **90**, 270 (1953)
GREENBERG, O. W., and A. MESSIAH: preprint (1963) (unpublished)
GUIRAGOSSIÁN, Z. G. T.: Phys. Rev. Letters **11**, 85 (1963)
HAIG, F. R., T. F. JORDAN, and A. J. MACFARLANE: University of Rochester Report URPA-1 (1963), unpublished
HALPERN, F. R.: Phys. Rev. Letters, **12**, 252 (1964)
HAMERMESH, M.: Group Theory. Reading, Mass.: Addison-Wesley, 1962
HAND, L. N., D. G. MILLER, and R. WILSON: Rev. Mod. Phys. **35**, 335 (1963)
HELLAND, J.: Differential cross sections for elastic scattering of positive pi mesons on protons in the energy region 500 to 1600 MeV. University of California Report UCRL-10378 (August 1962), see also Phys. Rev. Letters **10**, 27 (1963)
HENLEY, E., and B. JACOBSOHN: Phys. Rev. **128**, 1394 (1962)
HOFSTADTER, R.: Rev. Mod. Phys. **28**, 214 (1956)
HOLLADAY, W. G., and H. G. VON BAEYER: private communication (1963) (to be published)
HOMER, R. J., Q. H. Khan, W. K. McFARLANE, J. S. C. McKEE, A. W. O'DELL, L. RIDDIFORD, P. G. WILLIAMS: Physics Lett. **9**, 72 (1964).
HULTHÉN, L., and K. V. LAURIKAINEN: Rev. Mod. Phys. **23**, 1 (1951)
ISLAM, M. M., and R. PIÑON: Phys. Rev. Lett. **12**, 310 (1964)
ITZYKSON, C., and M. JACOB: Phys. Letters **3**, 153 (1963)
JACKSON, J. D.: Physics of Elementary Particles. Princeton: Princeton University Press 1958
— Elementary Particle Physics and Field Theory, S. Barshay, Ed. New York: W. A. Benjamin 1963
JACOB, M., and G. C. WICK: Ann. Physics **7**, 404 (1959)
JAUCH, J. M.: Helv. Phys. Acta **33**, 711 (1960)
JOST, R.: Helv. Phys. Acta **30**, 409 (1957)
KALBFLEISCH, G. R., L. W. ALVAREZ, A. BARBARO-GALTIERI, O. I. DAHL, P. EBERHARD, W. E. HUMPHREY, J. S. LINDSEY, D. W. MERRILL, J. J. MURRAY, A. RITTENBERG, R. R. ROSS, J. B. SHAFER, F. T. SHIVELY, D. M. SIEGEL, G. A. SMITH, and R. D. TRIPP: Phys. Rev. Letters **12**, 527 (1964)
KATZ, A., and H. J. LIPKIN: Phys. Letters **7**, 44 (1963)
KENNEY, V. P., and C. N. VITTITOE: Proc. Athens Topical Conf. on Recently Discovered Resonant Particles, Ohio University, Athens, Ohio (1963), p. 246
KIRZ, J., and D. H. MILLER: quoted by Abolins et al. (1963)
KOPELMAN, J., M. BLOCK, and C. SUN: Bull. Am. Phys. Soc. **9**, 34 (1964)

KUZNETSOV, YE. V., YA. SHALOMOV, A. GRASHIN, and YE. P. KUZNETSOV: Phys. Letters 1, 314 (1962)

KYCIA, T. F., and K. F. RILEY: Phys. Rev. Letters 10, 266 (1963)

LANDAU, L.: Nuclear Phys. 3, 127 (1957)

LATTES, C., C. OCCHIALINI, and C. POWELL: Nature 160, 453, 486 (1947)

LEE, T. D., and C. N. YANG: Phys. Rev. 104, 254 (1956)

— — Nuovo cimento 3, 749 (1956a)

— — Phys. Rev. 105, 1671 (1957)

— — Phys. Rev. 109, 1755 (1958)

LEE, Y. K., L. W. MO, and C. S. WU: Phys. Rev. Letters 10, 253 (1963)

LEE, Y. Y., B. P. ROE, D. SINCLAIR, and J. C. VAN DER VELDE: Phys. Rev. Letters 12, 342 (1964)

LEIPUNER, L., W. CHINOWSKY, R. E. CRITTENDEN, R. ADAIR, B. MUSGRAVE, and F. SHIVELY: Phys. Rev. 132, 2285 (1963)

LICHTENBERG, D. B.: Stanford Linear Accelerator Report SLAC 13 (March 1963), unpublished

— Phys. Letters 4, 73 (1963a)

— Proc. Athens Topical Conf. on Recently Discovered Resonant Particles, Ohio University, Athens, Ohio (1963b), p. 153

—, and G. C. SUMMERFIELD: Phys. Rev. 127, 1806 (1962)

LINDENBAUM, S., W. LOVE, J. NIEDERER, S. OZAKI, J. RUSSELL, and L. YUAN: Phys. Rev. Letters 7, 352 (1961)

LINDENFELD, P., A. SACHS, and J. STEINBERGER: Phys. Rev. 89, 531 (1953)

LITHERLAND, A. E.: Can. J. Phys. 39, 1245 (1961)

LÜDERS, G.: Ann. Physics 2, 1 (1957)

— Lectures on Field Theory and the Many Body Problem, E. R. Caianiello, Ed., New York: Academic Press 1961

—, and B. ZUMINO: Phys. Rev. 106, 385 (1957a)

— — Phys. Rev. 110, 1450 (1958)

LUERS, D., I. S. MITRA, W. J. WILLIS, S. S. YAMAMOTO: Phys. Rev. 133, B 1276 (1964)

LÜTJENS, G., and J. STEINBERGER: Phys. Rev. Letters 12, 517 (1964)

LYUBARSKII, G. YA.: Application of Group Theory in Physics. New York: Pergamon Press 1960

MAGLIĆ, B., L. ALVAREZ, A. ROSENFELD, and M. STEVENSON: Phys. Rev. Letters 7, 178 (1961)

MARSHAK, R. E., and E. C. G. SUDARSHAN: Phys. Rev. 109, 1860 (1958)

— — Introduction to Elementary Particle Physics. New York: Interscience Publishers 1961

MEISNER, G. W., R. L. GOLDEN, B. B. CRAWFORD, and F. S. CRAWFORD JR.: University of California Report UCRL-11018, Sept. 16, 1963 (unpublished); International Conference on Fundamental Aspects of Weak Interactions. Brookhaven (Sept. 1963)

MEYER, S., E. ANDERSON, E. BLESER, L. LEDERMAN, J. ROSEN, J. ROTHBERG, and I.-T. WANG: Phys. Rev. 132, 2693 (1963)

MICHEL, L.: Progr. in Cosmic Ray Phys., pp. 142—144. New York: Interscience 1952

MILLER, D., G. ALEXANDER, O. DAHL, L. JACOBS, G. KALBFLEISCH, and G. SMITH: Phys. Letters 5, 279 (1963)

MINAMI, S.: Progr. Theoret. Phys. (KYOTO) 11, 213 (1954)

— Progr. Theoret. Phys. (Kyoto) 30, 722 (1963)

MORPURGO, G.: Ann. Rev. Nuclear Sci. 11, 41 (1961)

— Phys. Rev. 131, 2205 (1963)

MOYER, B. J.: Rev. Mod. Phys. 33, 367 (1961)

MULLER, F., R. BIRGE, W. FOWLER, R. GOOD, W. HIRSCH, R. MATSEN, L. OSWALD, W. POWELL, H. WHITE, and O. PICCIONI: Phys. Rev. Letters 4, 418 (1960)

NAKANO, J., and K. NISHIJIMA: Progr. Theoret. Phys. (Kyoto) 10, 581 (1953)

NAMBU, Y.: Phys. Rev. 106, 1366 (1957)

NE'EMAN, Y.: Nuclear Phys. 26, 222 (1961)

NISHIJIMA, K.: Fundamental Particles. New York: W. A. Benjamin 1963

156

ODIAN, A.: Private communication (1962)
OKUBO, S.: Progr. Theoret. Phys. (Kyoto) **27**, 949 (1962)
—, and R. E. MARSHAK: Nuovo cimento **28**, 56 (1963)
OMNÈS, R., and M. FROISSART: Mandelstam Theory and Regge Poles. New York: W. A. Benjamin 1963
PAIS, A.: Phys. Rev. **86**, 663 (1952)
— Phys. Rev. **112**, 624 (1958)
—, and R. E. JOST: Phys. Rev. **87**, 871 (1952a)
PANOFSKY, W., R. AAMODT, and J. HADLEY: Phys. Rev. **81**, 565 (1951)
PAULI, W.: Niels Bohr and the Development of Physics, W. PAULI, Ed. New York: McGraw Hill 1955
PEASLEE, D. C.: Phys. Rev. **117**, 873 (1960)
PEIERLS, R. F.: Phys. Rev. **118**, 325 (1960)
PESHKIN, M.: Phys. Rev. **123**, 637 (1961)
— Phys. Rev. **129**, 1864 (1963)
— Phys. Rev. **133**, B 428 (1964)
PEVSNER, A., M. NUSSBAUM, C. RICHARDSON, R. STRAND, T. TOOHIG, R. KRAEMER, P. SCHLEIN, M. BLOCK, A. ENGLER, R. GESSAROLI, and C. MELTZER: Phys. Rev. Letters **7**, 421 (1961)
PJERROU, G., D. PROWSE, P. SCHLEIN, W. SLATER, D. STORK, and H. TICHO: Phys. Rev. Letters **9**, 114 (1962)
PUPPI, G.: Ann. Rev. Nuclear Sci. **13**, 287 (1963)
REGGE, T.: Nuovo cimento **14**, 951 (1959)
— Nuovo cimento **18**, 947 (1960)
RICHTER, B.: Phys. Rev. Letters **9**, 217 (1962)
ROCHESTER, G. D., and C. C. BUTLER: Nature **160**, 855 (1947)
ROMAN, P.: Theory of Elementary Particles. 2nd Edition. Amsterdam: North-Holland 1961
ROOS, M.: Rev. Mod. Phys. **35**, 314 (1963), Physics Letters **8**, 1 (1964), Nuclear Physics **52**, 1 (1964)
ROPER, L. D.: Phys. Rev. Letters **12**, 340 (1964)
ROSE, M. E.: Elementary Theory of Angular Momentum. New York: John Wiley and Sons 1957
ROSENBLUTH, M.: Phys. Rev. **79**, 615 (1950)
ROSENFELD, A. H.: Strongly Interacting Particles and Resonances. UCRL-10492, University of California Report (1962)
— Stanford (1963a)
—, and W. E. HUMPHREY: Ann. Rev. Nuclear Sci. **13**, 103 (1963)
ROSS, M.: Private communication (1964)
SACHS, R. G.: Phys. Rev. **126**, 2256 (1962)
— Phys. Rev. **129**, 2280 (1963)
SAKATA, S.: Progr. Theoret. Phys. (Kyoto) **16**, 686 (1956)
SAKURAI, J. J.: Nuovo cimento **7**, 649 (1958)
— Phys. Rev. **113**, 1679 (1959); **115**, 1304 (1959)
— Ann. Physics **11**, 1 (1960)
— Invariance Principles and Elementary Particles. Part II. University of Chicago Report EFINS 61—30 (June 1961)
— Lectures in Theoretical Physics, Brandeis Summer Institute. New York: W. A. Benjamin 1961a
— Phys. Rev. Letters **9**, 472 (1962)
SALAM, A.: Nuovo cimento **5**, 299 (1957)
SAMIOS, N., A. BACHMANN, R. LEA, T. KALOGEROPOULOS, and W. SHEPHARD: Phys. Rev. Letters **9**, 139 (1962)
SCHLEIN, P., W. SLATER, L. SMITH, D. STORK, and H. TICHO: Phys. Rev. Letters **10**, 368 (1963)
SCHWARTZ, M.: Phys. Rev. Letters **6**, 556 (1961)
SCHWEBER, S. S.: An Introduction to Quantum Field Theory. Evanston, Ill.: Row, Peterson, and Co. 1961
SCHWINGER, J.: Phys. Rev. **104**, 1164 (1956)
— Ann. Physics **2**, 407 (1957)

Selove, W., V. Hagopian, H. Brody, A. Baker, and E. Leboy: Phys. Rev. Letters 99, 272 (1962)

Shafer, J. B., J. J. Murray, and D. O. Huwe: Phys. Rev. Letters 10, 179 (1963)

Shaklee, F., G. Jensen, B. Roe, and D. Sinclair: Bull. Am. Phys. Soc. 9, 34 (1964)

Shirokov, M. I.: J. Exp. Theor. Phys. 41, 190 (1961) [Soviet Phys. JETP 14, 138 (1962)]

—, and E. O. Okonov: J. exp. Theor. Phys. 39, 285 (1960) [Soviet Phys: JETP 12, 204 (1961)]

Sodickson, L., M. Wahlig, I. Mannelli, D. Frisch, and O. Fackler: Phys. Rev. Letters 12, 485 (1964)

Stapp, H. P.: Phys. Rev. 125, 2139 (1962); 128, 1962 (1962)

Stevenson, M., L. Alvarez, B. Maglić, and A. Rosenfeld: Phys. Rev. 125, 687 (1962); 125, 2208 (1962)

Stonehill, D., C. Baltay, H. Courant, W. Fickinger, E. Fowler, H. Kraybill, J. Sandweiss, J. Sanford, and H. Taft: Phys. Rev. Letters 6, 624 (1961)

Taher-Zadeh, M., D. Prowse, P. Schlein, W. Slater, D. Stork, and H. Ticho: Phys. Rev. Letters 11, 470 (1963)

Tarjanne, P., and V. L. Teplitz: Phys. Rev. Letters 11, 447 (1963)

Treiman, S. B.: Phys. Rev. 101, 1216 (1956)

— Phys. Rev. 128, 1342 (1962a)

—, and C. N. Yang: Phys. Rev. Letters 8, 140 (1962)

Tripp, R., M. Watson, and M. Ferro-Luzzi: Phys. Rev. Letters 8, 175 (1962)

Wali, K. C.: Phys. Rev. Letters 9, 120 (1962)

Walker, W., J. Boyd, A. Erwin, P. Satterblom, M. Thompson, and E. West: Phys. Letters 8, 208 (1964)

Wangler, T., A. Erwin, and W. Walker: Bull. Am. Phys. Soc. 9, 34 (1964); Phys. Letters 9, 71 (1964)

Watson, M. B., M. Ferro-Luzzi, and R. D. Tripp: Phys. Rev. 131, 2248 (1963)

Weyl, H.: Theory of Groups and Quantum Mechanics. Princeton: Princeton University Press 1931

Wick, G., A. S. Wightman, and E. P. Wigner: Phys. Rev. 88, 100 (1952)

Wightman, A. S.: Thesis, Princeton University (1949)

— Phys. Rev. 77, 521 (1950)

Wigner, E.: Phys. Rev. 98, 145 (1955)

— Group Theory. New York: Academic Press 1959

Williams, W. S. C.: An Introduction to Elementary Particles. New York: Academic Press 1961

Willis, W., R. Adair, H. Courant, H. Filthuth, P. Franzini, R. Glasser, A. Minguzzi, A. Segar, R. Burnstein, T. Day, F. Hertz, B. Kehoe, M. Sakitt, N. Seeman, B. Sechi-Zorn, and G. Snow: Bull. Am. Phys. Soc. 8, 349 (1963)

Wilson, R.: The Nucleon-Nucleon Interaction. New York: Interscience Publishers, John Wiley 1963

Wojcicki, S. G., G. R. Kalbfleisch, and M. H. Alston: Phys. Letters 5, 283 (1963)

Wolf, S., N. Schmitz, L. Lloyd, W. Lasker, F. Crawford, J. Button, J. Anderson, and G. Alexander: Rev. Mod. Phys. 33, 439 (1961)

Wu, C., E. Ambler, R. Hayward, D. Hoppes, and R. Hudson: Phys. Rev. 105, 1413 (1957)

Yang, C. N.: Phys. Rev. 74, 764 (1948)

— Phys. Rev. 77, 242 (1950)

—, and R. L. Mills: Phys. Rev. 96, 191 (1954)

Zorn, B. Sechi: Phys. Rev. Letters 8, 282 (1962)

References for sections 29—33

ABASHIAN, A., R. ABRAMS, D. CARPENTER, G. FISHER, B. NEFKENS, and J. SMITH:
Phys. Rev. Letters 13, 243 (1964)
ABRAMS, G. et al.: Phys. Rev. Letters 13, 670 (1964)
ADERHOLZ, M. et al.: Phys. Letters 10, 226 (1964)
ALITTI, J. et al.: Phys. Letters 15, 69 (1965)
ALLARD, J. et al.: Phys. Letters 12, 143 (1964)
ANDERSON, J., F. CRAWFORD, R. GOLDEN, D. STERN, T. BINFORD, and V. LIND:
Phys. Rev. Letters 14, 475 (1965)
BADIER, J. et al.: International Conference on High Energy Physics (Dubna, 1964).
BAZIN, M., H. BLUMENFELD, W. NAUENBERG, E. SEIDLITZ, and C. CHANG: Phys.
Rev. Letters 14, 154 (1965)
BÉG, M., B. LEE, and A. PAIS: Phys. Rev. Letters 13, 514 (1964)
BELL, J., and J. PERRING: Phys. Rev. Letters 13, 348 (1964)
BERNARDINI, G. et al.: Phys. Letters 13, 86 (1964)
BERNSTEIN, J., N. CABIBBO, and T. D. LEE: Phys. Letters 12, 146 (1964)
BINGHAM, H. et al.: Phys. Letters 9, 201 (1964)
BLOCK, M. et al.: Phys. Letters 12, 281 (1964)
BLUDMAN, S.: preprint (1964)
BÖCK, R. et al.: Phys. Letters 12, 65 (1964)
BOWEN, T., D. DeLISE, R. KALBACH, and L. MORTARA: Phys. Rev. Letters 13,
728 (1964)
BROWN, L., and H. FAIER: Conference on Symmetry Principles at High Energy,
Coral Gables, Florida (1965)
CHARRIÈRE, G. et al.: Phys. Letters 15, 66 (1965)
CHRISTENSON, J., J. CRONIN, V. FITCH, and R. TURLAY: Phys. Rev. Letters 13,
138 (1964)
CHUNG, S. et al.: Phys. Rev. Letters 12, 621 (1964)
COOPER, W. et al.: Phys. Letters 8, 365 (1964)
DALITZ, R. H., and R. G. MOOREHOUSE: Phys. Letters 14, 159, 356 (1965)
DAUBER, P., W. SLATER, L. SMITH, D. STORK, and H. TICHO: Phys. Rev. Letters
13, 449 (1964)
DE BOUARD, X. et al.: Phys. Letters 15, 58 (1965)
DURAND, L., and Y. CHIU: Phys. Rev. Letters 14, 329 (1965)
EBERHARD, P. et al.: Phys. Rev. Letters 14, 466 (1965)
FEINBERG, G.: Third Eastern Theoretical Physics Conference (Oct. 1964)
FRANZINI, P., B. LEONTIĆ, D. RAHM, N. SAMIOS, and M. SCHWARTZ: Phys. Rev.
Letters 14, 196 (1965)
GALBRAITH, W. et al.: Phys. Rev. Letters 14, 383 (1965)
GELFAND, N. et al.: Phys. Rev. Letters 12, 567 (1964)
GELL-MANN, M.: Phys. Letters 8, 214 (1964)
GOLDBERG, M. et al.: Phys. Rev. Letters 12, 546 (1964a)
— Phys. Rev. Letters 13, 249 (1964b)
GOLDHABER, G.: Conference on Symmetry Principles at High Energy, Coral Gables,
Florida (1965)
GOTTFRIED, K., and J. D. JACKSON: Nuovo Cimento 34, 735 (1964)
GREINER, D., W. OSBORNE, and W. BARKAS: Phys. Rev. Letters 13, 284 (1964).
GÜRSEY, F., T. D. LEE, and M. NAUENBERG: Phys. Rev. 135, B467 (1964a)
—, and L. A. RADICATI: Phys. Rev. Letters 13, 299 (1964b)
HAGOPIAN, V., and W. SELOVE: Phys. Rev. Letters 10, 533 (1963)
HAQUE, N. et al.: Phys. Letters 14, 338 (1965)
HARDY, L., S. CHUNG, O. DAHL, R. HESS, J. KIRZ, and D. MILLER: Phys. Rev.
Letters 14, 401 (1965)
HÖHLER, G., and J. GIESECKE: Phys. Letters 13, 149 (1964)
HUWE, D. O.: Univ. of California report UCRL-11291, 1964 (unpublished)
KALBFLEISCH, G. et al.: Phys. Rev. Letters 12, 527 (1964a)
—, O. DAHL, and A. RITTENBERG: Phys. Rev. Letters 13, 349a (1964b)
KLEIN, M., and M. ROSS: 1965, unpublished

LANDER, R., M. ABOLINS, D. CARMONY, T. HENDRICKS, N. XUONG, and P. YAGER: Phys. Rev. Letters **13**, 346a (1964)

LEE, T. D.: Phys. Rev. **137**, B1620 (1965)

LEIPUNER, L., W. CHU, R. LARSEN, and R. ADAIR: Phys. Rev. Letters **12**, 423 (1964)

MATTHEWS, P. T., and A. SALAM: Phys. Rev. **115**, 1079 (1959)

MCINTURFF, A., and C. ROOS: Phys. Rev. Letters **13**, 246 (1964)

MILLER, D., A. KOVACS, R. MCILWAIN, T. PALFREY, and G. TAUTFEST: Phys. Letters **15**, 74 (1965)

MORRISON, D. R. O.: Phys. Letters **9**, 199 (1964)

OLSSON, M. G.: Phys. Rev. Letters **14**, 118 (1965)

PAN, Y., and R. ELY: Phys. Rev. Letters **13**, 277 (1964a)

PJERROU, G., P. SCHLEIN, W. SLATER, L. SMITH, D. STORK, and H. TICHO: Phys. Rev. Letters **14**, 275 (1965)

ROSENFELD, A., A. BARBARO-GALTIERI, W. BARKAS, P. BASTIEN, J. KIRZ, and M. ROOS: Rev. Mod. Phys. **36**, 977 (1964a)

ROSENFELD, A. H.: In: "Nucleon Structure", ed. by R. HOFSTADTER and L. I. SCHIFF, Stanford University Press (1964b), p. 155

SACHS, R. G.: Phys. Rev. Letters **13**, 286 (1964)

SAKITA, B.: Phys. Rev. **136**, B1756 (1964a)

— Phys. Rev. Letters **13**, 643 (1964b)

SMITH, G. et al.: Phys. Rev. Letters **13**, 61 (1964)

— J. LINDSEY, J. MURRAY, and J. SCHAFER: Phys. Rev. Letters **14**, 25 (1965).

SUNDERLAND, J. et al.: Bull. Am. Phys. Soc. **10**, 35 (1965)

WIGNER, E.: Phys. Rev. **51**, 106 (1937)

WU, T. T., and C. N. YANG: Phys. Rev. Letters **13**, 380 (1964)

ZWEIG, G.: CERN reports 8181/TH401 and 8419/TH412, 1964 (unpublished)